Water Management in the
Yellow River Basin of China

T0339352

Water Management in the Yellow River Basin of China

BY CHARLES GREER

University of Texas Press
Austin

Grateful acknowledgment is made to the following for permission
to reprint previously published material:
    The Water Information Center, Inc., 7 High Street, Huntington,
New York 11743, for Table 1-1, which originally appeared in *The
Water Encyclopedia* (1970).
    American Geographical Society, Broadway & 15th Street, New
York, New York 10032, for Table 6-1, which originally appeared in
Geographical Review 63, no. 1 (January 1973).

Library of Congress Cataloging in Publication Data

Greer, Charles, 1942–
    Water management in the Yellow River Basin of China.

    Bibliography: p.
    Includes index.
    1. Water resources development—China—Hwang-ho Valley.
I. Title.
TC502.H8G73   333.9′1′009511   78-15303

ISBN: 978-0-292-74104-1

Copyright © 1979 by the University of Texas Press

All rights reserved

First paperback printing 2012.

Contents

Illustrations

TABLES

Abbreviations Used

CB	*Current Background*, U.S. Consulate, Hong Kong
CCNW	*Ch'ang-ch'eng Nei Wai (Within and Without the Great Wall*, Newspaper), Kuang-chou
CIFRC	China International Famine Relief Commission
CKHW	*Chung-kuo Hsin-wen (China News Service)*
CKSL	*Chung-kuo Shui-li (China Water Conservancy)*
CNA	*China News Analysis*, Hong Kong
CNS	*China News Service (Chung-kuo Hsin-wen)*
CPG	Central Peoples' Government
CR	*China Reconstructs*, China Welfare Institute, Peking
ECMM	*Extracts from China Mainland Magazines*, U.S. Consulate, Hong Kong
FBIS	Foreign Broadcast Information Service, U.S. Department of State
HCJP	*Hsu-chou Jih-pao (Hsu-chou Daily News)*
HHC	*Huang-ho Chih (Yellow River Chronicle)*, Published by Yellow River Water Conservancy Commission, 1935
HHSW	*Huang-ho Shui-wen (Yellow River Hydrology)*, published by Yellow River Water Conservancy Commission, 1945
HKTKP	*Ta-kung Pao (Ta-kung News)*, Hong Kong
HKWHP	*Wen-hui Pao (Wen-hui News)*, Hong Kong
HTJP	*Hsing-tao Jih-pao (Hsing-tao Daily News)*, Hong Kong

JMJP	*Jen-min Jih-pao (Peoples' Daily)*, Peking
JPRS	Joint Publications Research Service, U.S. Department of Commerce, Washington, D.C.
KJJP	*Kung-jen Jih-pao (Workers' Daily)*, Peking
KMJP	*Kuang-ming Jih-pao (Kuang-ming Daily News)*, Peking
LYCCC	*Li Yi-chih Ch'uan-chi (The Collected Works of Li Yi-chih)*, Taipei, 1956
NCNA	New China News Agency (Hsin-hua She)
NMKJP	*Nei-meng-ku Jih-pao (Inner Mongolian Daily News)*, Huhehot
SCMM	*Survey of China Mainland Magazines*, U.S. Consulate, Hong Kong
SCMP	*Survey of the China Mainland Press*, U.S. Consulate, Hong Kong
SJP	*Shensi Jih-pao (Shensi Daily News)*, Sian
SLFT	*Shui-li Fa-tien (Water Conservancy and Electric Power Generation)*, Ministry of Water Conservancy and Electric Power
SLHC	*Shui-li Hsing-cheng (Water Conservancy Administration)*, Chinese Ministry of Water Conservancy, 1947
SLHP	*Shui-li Hsueh-pao (Journal of Hydraulic Engineering)*, Hydraulic Engineering Society of China
SLYTL	*Shui-li Yu Tien-li (Water Conservancy and Electric Power)*, Ministry of Water Conservancy and Electric Power
THJP	*Ta-hsiang Jih-pao (Ta-hsiang Daily News)*, Tsinan
TKP	*Ta-kung Pao (Ta-kung News)*, Peking
TLCS	*Ti-li Chih-shih (Geographic Knowledge)*, Geographic Society of China
TLHP	*Ti-li Hsueh-Pao (Acta Geographica Sinica)*, Geographic Society of China
WRPC	Water Resources Planning Commission, Ministry of Economics, Taipei

YCWP   *Yang-ch'eng Wan-pao (Yang-ch'eng Evening News)*, Kuang-chou

YRPS   *Yellow River Project Studies*, Public Works Commission, Supreme Economic Council, Nanking

YRWCC   Yellow River Water Conservancy Commission (Huang-ho Shui-li Wei-yuan-hui)

Introduction
When the River Runs Clear

Water resources management in the Yellow River basin reflects all the important aspects of general development under the Communist government in China. The most pressing problem—elimination of disastrous floods in the basin—has been solved, just as in agriculture the basic problem of feeding the population has been solved, and in industry the foundation of an industrial economy has been established. But, like the rates of agricultural mechanization and industrial growth, the pace of modernization in Yellow River management has been slower than that anticipated in the ambitious basin-wide development plan announced in 1955.

The slower pace of modernization has been intimately linked with the emergence in the last twenty years of a unique Chinese method of development. The method is characterized by "walking on two legs" (integration of modern and traditional techniques) and by self-reliance (both independence from foreign aid and local autonomy from central government control on small- and medium-sized projects). Implementing this method in Yellow River development has meant that features such as power generation capacity and total irrigated area have not reached the original ambitious targets. At the same time, however, it has meant that progress is achieved at relatively low cost to the central government, and without technical or economic dependence on a foreign power.

The uniqueness of the Chinese method has attracted great attention among China specialists around the world. Much recent scholarship has focused on such topics as the evolution and operation of the commune system, the nature and performance of China's economic institutions, and the role of party and ideology in modernization efforts. While it contributes much to our understanding of how contemporary Chinese society operates, such focus has limitations as well. It makes difficult any effort to see the Chinese experience in a broader perspective. For example, it does not ask what parts of contemporary uniqueness may have roots in pre-Communist Chinese society; it tends to obscure such continuities as do exist across the unquestionably dramatic hiatus that 1949 does represent.

Nor does such a focus allow for easy comparison with other countries at a similar stage in the transformation to industrial society. The literature on economic development shows the growing sentiment for independence from foreign domination in the modernization process, but China—undoubtedly the best example of such independence—often is treated superficially in this literature because Chinese development experience has been insufficiently interpreted to an audience wider than that of the Chinese studies community.

Any study of modernization in Yellow River management must of necessity take a larger perspective. This process clearly spans the 1949 revolution, for example. Foreign interest in Yellow River management began in the late nineteenth century and culminated in the Soviet assistance decade of the 1950s. Within this long period two distinct subperiods are crucial for understanding how China has emerged with a great deal of independence in recent decades. The subperiods are the late Republican era of the 1930s–1940s and the latter half of the 1960s. In the former, China clearly was acquiring Western hydraulic science and technology on her own terms to a greater degree than has been possible for any other developing country. In the latter, decisions were made to revitalize traditional water management methods (which had not been supplanted by foreign methods as they have been so often elsewhere in the world) to augment modernization in place of continued foreign aid. Although this study does not treat in detail how such a program might be adapted to specific conditions in other countries, it is hoped that it will make clear how the process was possible in the Yellow River basin.

This study also is concerned with the problems of mutual adaptation between society and habitat. Terms such as "modernization" and "development" refer of course to the transition from agrarian society to industrial society. To understand this transition in the Yellow River basin requires a focus on this mutual adaptation and how it changes. The physical parameters of the specific resource—the Yellow River system—must be considered, as must the adaptation made by traditional society, before the innovations made in the past century can be understood with any success.

Nor can attention be limited to technological innovation in this adaptation. Organizational and ideological innovation, equally important components of society in adaptation, are given equal emphasis in this study, and the relationship between the three types of innovation is considered. The Chinese ability to substitute organizational innovation for technological innovation and to use new ideological forms as a motivational tool are nowhere better illustrated than in the Yellow River development scheme.

The framework utilized to discuss so large a topic in a single short book is that provided by Gilbert White.[1] White asks three questions of any given water resources development project: what are the purposes of development; what are the means of development; and what are the agencies involved in development. Asking these questions of historical Yellow River projects helps to extract their essential elements for comparison with those of modern development. Asking them of recent projects provides a framework for understanding the interrelationship of technological, organizational, and ideological aspects of the development scheme.

The materials available for such a study, while not complete by any means, are satisfactory by the standards of information generally available from China. Not only has river development been more fully reported in the press and discussed in technical journals than many other kinds of development, but the wealth of material on Yellow River management from traditional China and certain documents from the Republican period provide sufficient information for undertaking the inquiry outlined above.

This study is therefore able to proceed on the following line of questioning: what are the specific river management problems in the Yellow River basin and how are they being solved by the modern basin-wide scheme; how is the unique Chinese development model illlustrated in these solutions; to what degree is river management based on continuity with traditional Chinese practices, and in what ways have foreign influences been incorporated; does the evidence indicate the Chinese are beginning a truly new form of river management within industrial society, or have the past three decades represented instead a transition phase from which river management will evolve toward patterns already established elsewhere in the world?

To answer these questions the study begins with a description of the specific problems of Yellow River management, focusing on silt control as the root cause of other management problems. It then summarizes historical management experience and discusses the impact of foreign river management strategies through the decade of the 1950s. The importance of traditional Chinese methods in recent projects is then discussed, and finally an evaluation of the Chinese strategy is provided.

The degree to which the Yellow River dominated life in north China historically is reflected not only in the river's name, which refers to the eternal silt burden, but also in the idiom "when the river runs clear," a phrase used to mean "never," which expresses the perceived impossibility of the silt problem's ever being solved. At times in the past three decades, using the same enthusiasm with which they speak of "the foolish old man who moved the mountain," Chinese publications have referred to making the river run clear for the first time in history. While this goal is only marginally closer to being realized now than it has been in the past, the way it is incorporated in the modern effort is singular. Like the poems of Mao Tse-tung, which are revolutionary in content but very traditional in style, the Yellow River development program—and all development in China—must be studied from a perspective which both penetrates its uniqueness and interprets its significance for the world at large.

Water Management in the
Yellow River Basin of China

Chapter 1
The Problem of
Yellow River Management

The Yellow River, or Huang Ho, is more widely known than many other rivers in the world that are substantially larger.[1] One reason for this is that the Yellow River basin was the "cradle of Chinese civilization," and it has remained an integral part of the Chinese national territory for more than three thousand years. Another reason is the terrible destruction caused by the river's periodic floods. Throughout history the river has repeatedly broken through its levees to rampage over the densely populated North China Plain. The tremendous loss of life and property and the extreme misery caused by such floods have earned the river the epithet "China's Sorrow."

The Yellow River floods arose from a mixture of human and natural causes. Artificial channels across the North China Plain and erosion in the middle portion of the basin are features for which human society is partially responsible and which contribute to the severity of the flood problem. Long before these human impacts were first made, however, certain hydrologic and geomorphic processes were present in the basin which created conditions that were less than favorable for dense human settlement. These natural processes give rise to drought and flood conditions, as well as to siltation of the river channel—the roots of Yellow River management difficulties.

Natural Processes in the Basin

Two features of the evolution of the Yellow River drainage pattern during recent geologic history bear directly on modern management problems. These features are stream capture and eustatic warping of the earth's crust. The large northern loop in the middle

Table 1-1. *Major Rivers of the World*

River	Length (mi.)	Drainage Area (1,000 sq. mi.)	Average Discharge at Mouth (1,000 cfs)
Amazon	3,900	2,231	7,500
Congo	2,900	1,550	1,400
Yangtze	3,100	750	770
Brahmaputra	1,680	361	700
Ganges	1,540	409	660
Yenisei	2,800	1,000	614
Mississippi	3,988	1,244	611
Orinoco	1,700	340	600
Lena	2,800	936	547
Parana	2,450	890	526
St. Lawrence	1,900	498	500
Irrawaddy	1,250	166	479
Ob	3,200	959	441
Mekong	2,500	310	390
Amur	2,900	712	388
Tocantins	1,700	350	360
Mackenzie	2,525	697	280
Magdalena	950	93	265
Columbia	1,214	258	256
Zambezi	1,600	500	250
Danube	1,725	315	218
Niger	2,600	430	215
Indus	1,700	358	196
Yukon	1,800	360	180
Pechora		126	144
Uruguay	1,000	90	136
Kolyma		249	134
Si Kiang	1,650	46	127
Godavari		115	127
Dvina	400	139	124
Yellow River	2,700	260	116
Frazer	695	92	113
Nile	4,000	1,150	100

Source: David Keith Todd, *The Water Encyclopedia*, pp. 118–121.
Note: No reliable figure for the discharge of the Yellow River at or near its mouth is available. The figure of 116,000 cubic feet per second (3,238 cubic meters per second) is much too high. The flow at Shan-hsien, 700 km from the mouth, is less than half that amount (1,375 cms), and although a few small tributaries join the main stream below Shan-hsien, the discharge of the Yellow River certainly is not doubled as it crosses the flood plain. The rank of the Yellow River in this table, therefore, should be lower than thirty-first, but exactly where it ranks cannot be established without a definite value for average discharge at the mouth.

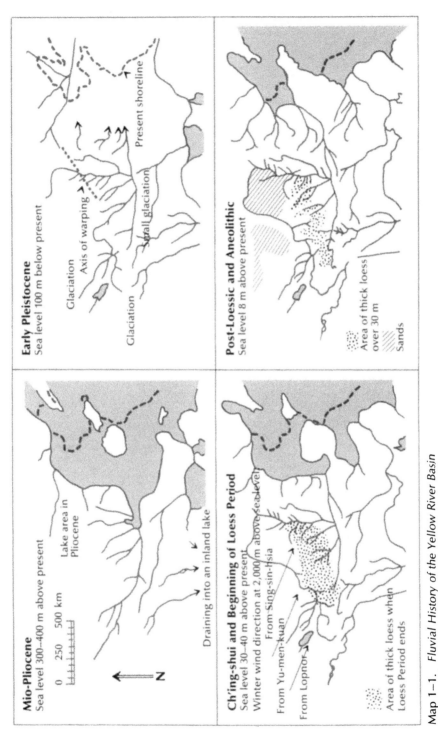

Map 1–1. *Fluvial History of the Yellow River Basin*
Source: After Ting et al. 1946.

portion of the basin evolved through a process of stream capture during Pleistocene times (Map 1-1). The northern loop accounts for both the aridity in the middle part of the basin and the knickpoint in the Yellow River's gradient. Because of the aridity, much of the runoff from the headwaters of the river is lost in the middle basin, and drought conditions are much more severe downstream than they would be if the river did not traverse the xeric environment of the northern loop.

The knickpoint is located near T'o-k'e-t'o east of Pao-t'ou, where the river turns south along the border between Shansi and Shensi provinces (Table 1-2). Geologically this is the place where the river leaves the edge of the Pre-Cambrian platform which is overlain by Quarternary alluvium and cuts through the loess mantle that covers Mesozoic structures.[2] The knickpoint is not especially significant in itself, but because the gradient becomes steeper across the Loess Plateau, the river acquires a heavier load of silt than it would if the gradient were smooth across the Loess Plateau.

Eustatic warping of the earth's crust, coupled with the phases of continental glaciation, accounts for various sea level fluctuations on the North China Plain. S. Ting et al. describe the following fluctuations: a level three to four hundred meters higher than the present

Table 1-2. *Gradient of the Yellow River Mainstream*

Section	Average Drop (meters/km)	Average Slope
Source to Ch'ing-t'ung Gorge	1.61	1:621
Ch'ing-t'ung Gorge to T'o-k'e-t'o	.18	1:5555
T'o-k'e-t'o to T'ung-kuan	.72	1:1389
T'ung-kuan to Meng-chin	.76	1:1316
Meng-chin to mouth	.16	1:6250

Source: Chang Shichin, *General Description of the Yellow River Basin.*

level during the Pliocene which entirely submerged the river below T'ung-kuan; a level as much as one hundred meters below the present level in Lower Pleistocene times, associated with the Lung Shan uplift and a glacial period contemporary with the Riss in Europe; a level about eight meters above the present level at the end of the last glaciation; and a subsequent adjustment to the present level causing perceptible incision of the river channel upstream.[3]

Contemporary eustatic movement of the North China Plain is of fundamental importance to human use of the lower Yellow River basin. If eustatic movement were to raise the plain slightly, more land would emerge as the shore line shifted seaward. This might seem to be what actually is happening since the Yellow River throughout history has extended its delta by several meters per year. But Feng suggests that this extension is merely the action of huge amounts of sediment deposition overcompensating for a counteracting eustatic subsidence.[4] His suggestion is supported by the fact that the rivers on the plain are not incising their channels, but instead range freely over the plain, even with the construction of dike-restraining works by the Chinese throughout history. As Feng points out, such action by the Hai, Yellow, and Huai rivers has served in recent geologic times to build up the entire plain that now connects the Ta-pieh, Fu-niu, T'ai-hang, and other mountain ranges with the T'ai-shan mountain outlier of Shantung, formerly an island.

The geologic role performed by the Yellow River, therefore, is the removal of material from the western uplands and the deposition of that material on the slowly subsiding plain. To accomplish this, the river must range freely over the length and breadth of the area of deposition. The difficulties presented by this process for permanent human settlement are obvious, and they are exacerbated by the presence of loess in the middle portion of the basin.

Before the silt, drought, and flood problems are examined in more detail, it will be useful to summarize the general hydrological characteristics and related water management problems in various subregions of the basin. Five major subregions may be identified for this purpose: the Chinghai Upland, the Ninghsia–Great Bend area, the Loess area, the Shansi-Honan mountains, and the Yellow River Plain (Map 1-2).

The Chinghai Plateau, including the T'ao and Huang-shui tributaries, comprises nearly one-third of the basin catchment area

(Map 1-2 and Table 1-3), yet it contributes two-thirds of the annual runoff of the river. The steep rock slopes, low evaporation, and high retention of moisture in the form of snow help to account for a runoff coefficient on the plateau of more than 50 fifty percent.[5] Equally important, and in sharp contrast to other subregions, is the relatively even distribution of runoff over the year (Table 1-4), reflected in the ratio of minimum discharge to maximum discharge of 1 : 25.

The major water resource potentials of this section lie in storage and hydropower development in the gorges through which the river drops west of Lanchou. Between Kuei-te and Lanchou, a distance of about three hundred kilometers, fully one-fourth of the river's length is occupied by gorges one hundred meters or less in width. In the principal tributaries above Lanchou, irrigation and reclamation offer considerable potential benefits to agricultural production. The Huang-shui and T'ao basins have hydrologic regimes similar to that of the upper Yellow River, but lower elevations along these tributaries are milder climatically than the headwaters region, representing transition zones between the high plateau and the arid lands to the north and east.

Aridity is the main feature of the Ninghsia–Great Bend section of the basin, which extends from north of Lanchou to T'o-k'e-t'o. Very few tributaries join the Yellow River in Ninghsia or along the Great Bend (Ho-t'ao), as the top of the northern loop of the river is called. Annual precipitation decreases to about 120 millimeters (5 inches) (Map 1-3), evaporation rates are several times higher than precipitation rates (Table 1-5), and the river is bounded by deserts with intermittent streams and playa lakes. One result of these conditions is that the average annual discharge at Pao-t'ou is lower than that at Lanchou (Table 1-4).

Between Lanchou and southern Ninghsia six major gorges constrict the river, thus creating a significant potential for hydroelectric power generation. Irrigation and reclamation are the major tasks on the Ninghsia and Ho-t'ao plains, however. Alluvial soils are present along this section where the river's gradient decreases, but so are sandy and alkaline soils. Much water is lost through filtration into the sand, and a braided channel tendency of the main stream is accentuated by the construction of drainage ditches and irrigation canals. The potential benefits to agriculture of irrigation and reclamation work in such an environment are very great.

As the Yellow River flows southward between Shansi and Shensi

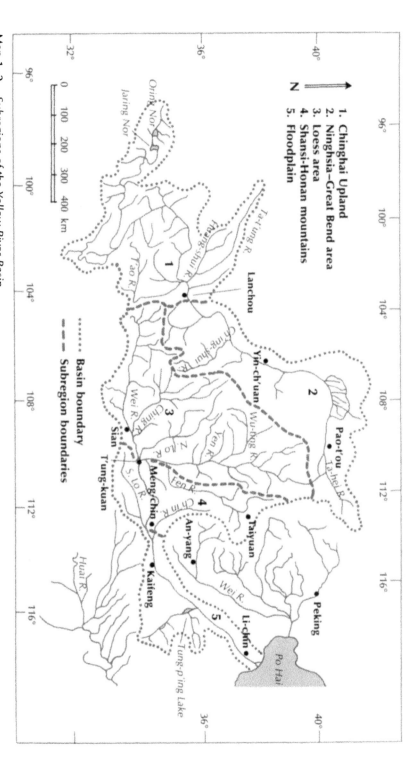

Map 1–2. *Subregions of the Yellow River Basin*

Table 1-3. *River Length and Drainage Area for Major Portions of the Basin*

Length (km) / Area (km²), measured between stations (from upstream station → downstream station).

From → / To ↓	Source	Lanchou	Ninghsia	Pao-t'ou	Lung-men	T'ung-kuan	Shen-hsien	Lo-k'ou
Lanchou	1,477.5 / 216,180							
Ninghsia	2,007.5 / 281,500	530.0 / 65,320						
Pao-t'ou	2,565.0 / 394,780	1,087.5 / 178,600	557.5 / 113,280					
Lung-men	3,475.0 / 515,378	1,997.5 / 299,198	1,467.5 / 233,878	910.0 / 120,598				
T'ung-kuan	3,603.0 / 712,588	2,125.5 / 496,408	1,595.5 / 431,088	1,038.0 / 317,808	128.0 / 197,210			
Shen-hsien	3,693.0 / 715,184	2,215.5 / 499,004	1,685.5 / 433,684	1,126.0 / 320,404	218.0 / 199,806	90.0 / 2,596		
Lo-k'ou	4,404.0 / 770,800	2,926.5 / 554,620	2,396.5 / 489,300	1,839.0 / 376,020	929.0 / 255,422	801.0 / 58,212	711.0 / 55,616	
Mouth	4,635.0 / 771,574	3,157.5 / 555,394	2,627.5 / 490,074	2,070.0 / 376,794	1,160.0 / 256,198	1,032.0 / 58,986	942.0 / 56,390	231.0 / 774

Source: Chang Shichin, *General Description.*

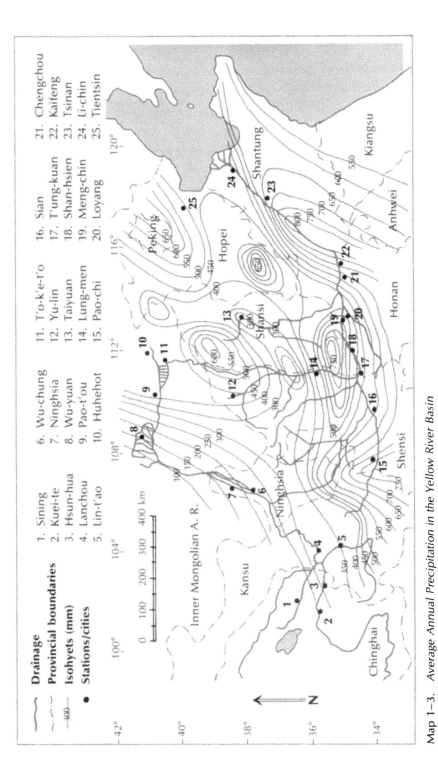

Map 1–3. *Average Annual Precipitation in the Yellow River Basin*
Source: After Ching Han-ying, ed. 1947.

Table 1-4. Discharge and Annual Runoff at Major Stations on the Yellow River Mainstream

Station	Discharge (cms)			Annual Runoff (billion cu. m)		
	Mean	Max.	Min.	Mean	Max.	Min.
Lanchou	1,105	5,750	290	31.4	43.9	23.8
Wu-chung		5,230	290			
Pao-t'ou	818	3,060	175			
Lung-men		19,750	134	41.3	50.4	29.2
Shan-hsien	1,375	29,000	140	54.4	72.3	38.4

Sources: Chang Han-ying, ed., Report on an Investigation of the Upper and Middle Reaches of the Yellow River (Huang-ho Shang-chung-yu K'ao-ch'a Pao-kao); YRWCC; V. T. Zaychikov, "Inland Waters," in idem, ed., The Physical Geography of China; Shen Yi, Purpose and Scope of the Yellow River Project Studies; HHSW.
Note: The table is based on data collected for only a few years and appearing in widely different sources, so the figures can only be taken as approximations.

Table 1-5. General Climatic Characteristics

Station	Mean January Temp. (Deg. C.)	Mean July Temp. (Deg. C.)	Av. Annual Precipitation (mm) (in.)		Av. Annual Evaporation (mm)
Headwaters	−16	−8	250	9.8	
Lanchou	− 6.8	22.8	325.7	12.8	1355.1
Sining	− 6.4	18.3	376	14.8	
Wu-chung	− 7.7	23.3	236.2	9.3	1511.3
Pao-t'ou	−14.1	22.2	409.5	15.7	1409.0
Lung-men	− 0.5	28.2	484.6	19.1	1491.0
Sian	− 0.8	28.1	535.3	21.1	1213.7
T'ung-kuan	− 2.4	26.9	471.1	18.6	1423.4
Taiyuan	− 7.3	25.4	370	14.6	
Shan-hsien	0.5	28.2	545.2	21.5	1249.4
Meng-chin			526.2	20.7	1617.6
Li-chin	− 3.2	27.6	487.5	19.2	1410.3

Sources: Chang Han-ying, ed., Report on an Investigation; G. N. Vitvitskiy, "Climate," in Zaychikov, ed., Physical Geography.

Table 1-6. *Size and Discharge of Tributary Basins*

River	Length (km)	Basin Area (sq km)	Discharge (cms) Mean	Max.	Min.
T'ao	230	29,678			
Huang-shui		30,320		860	12
Ta-t'ung	288			360	16
Ch'ing-shui		16,768			
Wu-ting	489	19,720	30	1,232	13
Yen	230				
Fen	550	38,000	40	1,950	0.5
Su		5,320			
Wei	670	144,760	300	5,880	18
Ching	443	58,930		8,451	0.4
N. Lo		27,020	30	2,050	
S. Lo		13,028		7,740	5.0
Yi		4,960			
Ch'in		10,500		2,740	1.0

Sources: Chang Shichin, *General Description*; Zaychikov, ed., *Physical Geography*; Chang Han-ying, ed., *Report on an Investigation*; HHC; W. H. Huang, *Soil and Water Conservation of the Yellow River Basin*; NCNA, Yenan, August 8, 1956, "Survey of Wu-ting River in Shensi Completed," *SCMP*, No. 1348 (August 13, 1956), p. 13; NCNA, Chengchou, October 30, 1956, "Ching River Survey Completed," *SCMP*, No. 1403 (November 2, 1956), p. 24.

Note: Discharge data have been collected for only a few years and appear in a variety of publications, so the figures are rough.

provinces, precipitation amounts increase, and tributaries once again become numerous (Table 1-6). Average annual precipitation is less than twenty inches over most of the Loess Plateau (Map 1-3), so irrigation remains fundamental to successful agricultural production, while silt becomes a major problem of river management. The Wu-ting, Yen, Northern Lo, Wei, and many lesser tributaries all have long-established irrigation works and chronic silt problems. North of T'ung-kuan, for example, the channel of the Yellow River has shifted ten kilometers eastward in the past three decades. This has made the Northern Lo River, which previously emptied into the Yellow River, a tributary of the lower Wei River. At the same time, the mouth of the Wei has been extended, thereby decreasing its gradient and accentuating the backwater effect the Yellow River exerts on the Wei in high water periods.[6]

After turning east at T'ung-kuan, the Yellow River once again descends through a constricted valley, with few tributaries and several gorges, before debouching on the North China Plain. The gorges provide key sites for hydropower generation and flood protection of the lower reaches. The tributaries which drain these Shansi-Honan mountains provide good development sites as well. The Ch'in River drains the T'ai-hang Shan of eastern Shansi, and the Southern Lo, with its branch the Yi River, drains the mountains of western Honan. These relatively small basins receive more precipitation than the ones farther west, and therefore contribute significant amounts of runoff annually to the mainstream. Key sites exist in each of these basins for irrigation and flood control development. The inclusion of the Fen and Su rivers of southwest Shansi in this subregion is based on their drainage characteristics, in spite of the fact that they enter the Yellow River above T'ung-kuan. Both drain mountainous regions rather than regions of primary loess, and their runoff and discharge characteristics are more similar to those of the tributaries farther east than to those of the rivers of the Loess Plateau.

No major tributaries enter the Yellow River between Chengchou and the Po Hai. Historically the primary task of water management on the lower course was flood control. Massive levees were constructed to stabilize the channel and to restrict the tendency of the river to change its course. These levees made diversion for irrigation or other uses very difficult. The potential benefit of such diversions is great in the unpredictable environment of the North China Plain, however, and with the solution of the flood problem they can be undertaken with increasing frequency and reliability.

Drought and Flood

The severity and complexity of specific water management problems in the Yellow River basin can be illustrated by closer examination of certain hydrologic characteristics. The root of the drought and flood problems lies in the erratic precipitation regime, which produces conditions marginal to unirrigated agriculture throughout much of the basin. Mean annual precipitation for the basin as a whole is 400 millimeters, or about 16 inches, but it is very unevenly distributed in both space and time. It ranges from 800 millimeters in western Shantung to about 120 millimeters in the Ho-t'ao region (Map 1-3). Parts of western Shansi and the Wei valley receive 600 millimeters or more, but most of the Loess Plateau receives less than 500 millimeters.

Five hundred millimeters (19.5 inches) is certainly enough to sustain agriculture if it is reasonably distributed throughout the year. In the Yellow River basin, however, 50 to 60 percent of total rainfall typically comes in three months, June through August (Table 1-7). The result is a surplus of summer rain and an extreme lack of moisture the rest of the year.

A reliable pattern of agriculture might be possible even in these conditions, if precipitation from year to year were relatively constant; but variation in annual rainfall amounts is also great, and this constitutes the most serious cause of drought conditions. Although average annual precipitation in the middle part of the basin is 470 millimeters, for example, one year in ten has a rainfall of less than 250 millimeters. At Pao-t'ou the yearly average from 1919 to 1932 was 320 millimeters, but only 180 millimeters fell in 1928 and 203 millimeters in 1930.[7] The fluctuation is most dramatic at Taiyuan, where the yearly total averages 370 millimeters, but dry years average 50 millimeters and wet years 700 millimeters.[8]

These conditions, coupled with the low water retention capacity of loess and other soils in the basin, and the fact that 35 percent of precipitation typically comes in sudden downpours, are responsible for the severe droughts which have affected large areas of north China throughout Chinese history. During the 1920s famines caused by such droughts often resulted in half the population of a given locality starving to death and most of the rest fleeing their homes in search of food.[9]

The Yellow River floods arise from the same set of basic environmental conditions that cause droughts. Although the amount of runoff in the Yellow River basin is relatively small (one-twentieth that of the Yangtze in a basin two-fifths as large), the seasonal concentration of runoff makes flooding a severe problem. The variation in mean monthly discharge is illustrated in Table 1-8. The sudden downpour of rainfall during the summer high water period, moreover, can bring a flood peak in a very short time. In mid-July 1958, the discharge at Hua-yuan-k'ou near Chengchou rose from 7,200 to 21,000 cubic meters per second in a period of twenty-five hours.[10] This flood crest, the highest since the disastrous flood of 1933, was caused by very heavy rainfall in the Ch'in and Southern Lo valleys, while precipitation in the rest of the basin was about

Table 1-7. *Precipitation Distribution by Seasons*

Station	March–May (mm)	(%)	June–August (mm)	(%)
Lanchou	49.3	15.2	196.2	60.2
Sining	56	12	211	56
Wu-chung	43.8	18.5	25.4	53.1
Wu-yuan	14.9	12.6	65.8	54.6
Huhehot	53.8	13.6	286.6	67.0
Yu-lin	39.1	8.7	308.2	68.3
Lung-men	77.5	16.2	268.7	55.2
T'ung-kuan	74.9	16.0	255.3	54.1
Sian	111.5	20.8	235.7	44.0
Shan-hsien	74.0	13.5	311.6	57.2
Meng-chin	63.7	12.1	340.4	64.7
Kaifeng	119.3	17.9	368.1	57.8
Li-chin	68.9	14.2	310.0	63.6

Sources: Chang Han-ying, ed., *Report on an Investigation*; Vitvitsky, "Climate," in Zaychikov, ed., *Physical Geography*.

Table 1-8. *Comparison of Monthly Discharge and Silt Content at Meng-chin*

Month	Mean Discharge (cms)	Mean Content of Dry Residue (%)
Jan.	300	0.4
Feb.	500	0.6
Mar.	700	0.8
Apr.	700	1.0
May	800	1.5
June	800	1.5
July	2,000	3.0
Aug.	3,500	5.0
Sept.	2,800	3.0
Oct.	1,500	2.0
Nov.	1,000	1.5
Dec.	600	1.5

Source: Zaychikov, ed., *Physical Geography*, p. 197.

September–November		December–February		Total	
(mm)	(%)	(mm)	(%)	(mm)	(in.)
71.5	22.0	8.7	2.6	325.7	12.8
105	29	4	3	376	14.8
55	23.3	12.0	5.1	136.2	5.4
38.2	31.8	1.2	1.0	120.1	4.4
79.9	18.6	5.1	1.2	425.4	16.7
95.5	21.5	6.9	1.5	449.7	17.3
111.1	23.0	27.3	5.6	484.6	19.1
125.9	26.2	15.9	3.2	472	18.6
171.0	32.0	17.1	3.2	535.3	21.1
143.9	26.4	15.7	2.9	545.2	21.5
101.4	19.2	21.3	4.0	526.8	20.7
125.7	18.8	37.7	5.6	650.8	25.6
84.3	17.3	24.3	4.9	487.5	19.2

normal. Table 1-9 illustrates the same point for other sections of the basin in earlier floods—a downpour in only one portion of the basin can cause flooding on the lower course.

The variability of total runoff from year to year further contributes to the severity of inundations in some years. Table 1-4 shows that although the average annual runoff at Shan-hsien from 1919 to 1953 was 54 billion cubic meters, the maximum was 72 billion and minimum was about half of that.

These conditions make it very evident that both irrigation and flood control are major tasks of any water management undertaking in the Yellow River basin. These problems are difficult enough, but efforts to find solutions are complicated even more by the huge amounts of silt in the river system.

Erosion and the Silt Problem

Certain questions remain unanswered as to the origin and deposition period of the loess, which forms a layer up to three hundred meters thick across northwest Shansi, Shensi, and eastern Kansu,

Table 1-9. *Origins of Flood Waters at Shan-hsien*
(Discharge in Cubic Meters per Second)

Date	Oct. 2, 1933	Aug. 7, 1935	Aug. 1, 1937	Aug. 4, 1942
From above Pao-t'ou	2,200	1,820	2,650	1,200
From between Pao-t'ou and Lung-men		4,600	3,345	21,800
From the Fen River	1,800	83	30	
Wei River	4,000	560	490	130
Ching River	12,000	1,420	20	290
Northern Lo River	300	125	290	100
Other	2,300	3,892	1,075	5,160[a]
From between T'ung-kuan and Shan-hsien		5,760	8,600	750
Total flood at Shan-hsien	22,600	18,260	16,500	29,430

Source: HHSW.
[a] Includes Fen River.

with lesser thickness over a large part of the Huang Plain. The source area generally is agreed to be centered in the arid regions of modern Sinkiang and Mongolia, and deposition probably began at least 45,000 years ago.[11] In spite of unanswered questions, however, there is no disagreement about the role the loess plays in the basin. The loose-packed loess, which covers approximately one-third of the basin area, is very easily eroded. Tributaries, down to the smallest gullies and rivulets, erode the loess and carry it away as suspended or bed load material.

Approximately one and one-half billion tons of loess are eroded annually in the Yellow River basin (Table 1-10). Half of that amount settles out of suspension as the river slows down across the flood-plain, and half of it is carried to the sea. As much as one centimeter per year is stripped from certain tributary basins in northern Shensi where erosion is most severe (Table 1-11 and Map 1-4). The equivalent of a layer three centimeters thick over the entire basin has been eroded in the last one hundred years, but as Table 1-12 shows, nearly 90 percent of the silt burden is acquired between Pao-t'ou and Shan-hsien. One-third of it comes from the Wei basin alone.

These conditions contribute to the heaviest silt load of any of the world's major rivers. In comparison to the average of thirty-four kilograms per cubic meter of suspended material in the Yellow

Table 1-10. *Annual Volume of Silt Runoff*

Station or Area	Average Annual Runoff (millions of tons)
Shan-hsien	
maximum	1,380
minimum	4,427 (1933)
August average	320
January average	532
Lanchou	216
Ching-shui Basin	60
Pao-t'ou	170
Between Pao-t'ou and Shan-hsien	1,200
Shansi Prov.	138
Wu-ting Basin	275
Wei Basin	476
Ching Basin	220
South Lo/Yi Basin	20
Ch'in Basin	14

Sources: Kuo Ching-hui, "The Silt of the Yellow River and Its Erosive Action (Huang-ho ti Ni-sha Chi Ch'i ch'in-shih Tzo-yung)," *TLCS*, August 1956, pp. 389–392; Zaychikov, ed., *Physical Geography*; NCNA, Taiyuan, July 22, 1955, "Soil Erosion along the Yellow River in Shansi," *SCMP*, No. 1095 (July 23, 1955), p. 41; NCNA, Yenan, August 8, 1956, "Survey of Wu-ting Completed"; NCNA, Cheng-chou, October 30, 1956, "Ching Survey Completed"; NCNA, Yin-ch'uan, November 22, 1963, "Antierosion Work Transforms Loess Hills in Northwest China," *SCMP*, No. 3109 (November 29, 1963), p. 13.

Note: The apparent discrepancies in the table arise from the following causes: different sources were used to obtain the figures, silt content figures refer to suspended load and do not take bed load into account, and considerable deposition takes place between Lanchou and Pao-t'ou.

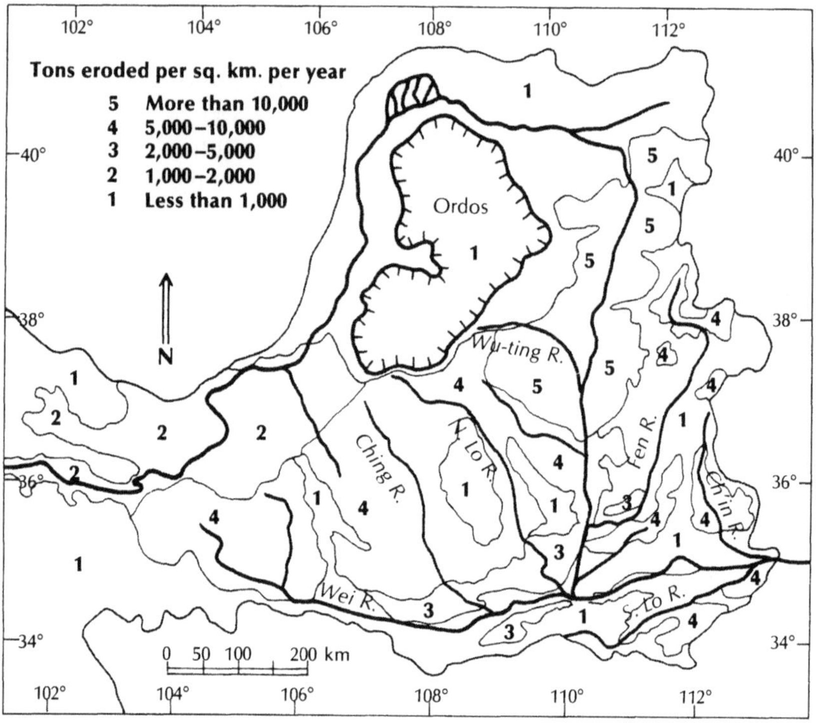

Map 1–4.   *Severity of Erosion in the Middle Yellow River Basin*
Source: After Kuo Ching-hui.

River at Shan-hsien, the Nile carries one kilo and the Colorado River carries ten kilos.[12] This tremendous silt load speeds up the river's alluvial action on the floodplain, so that the river must change its course much more frequently than it would with a smaller silt load.

The relation between the river's discharge and the amount of silt in suspension is critical to understanding how silt exacerbates the flood problem. Table 1-8 shows that the amount of silt per cubic meter of water increases as the discharge increases. Therefore, not only is there much more water in the river during flood periods, but there is also much more silt per unit measure of water. The amount of silt in suspension reaches its extraordinary maximums at periods of highest runoff (see Table 1-11).

Table 1-11. *Silt Content of the Yellow River and Major Tributaries*

Station/Tributary	Average Silt Content	
	Kilo/Cu M	% Weight
Hsun-hua	2	
Lanchou	3	
T'o-k'e-t'o	6	
Lung-men	28	
Shan-hsien	34	3.4
maximum recorded, Aug. '42	575	42.29
Ching River	161	
maximum recorded	978	
Wu-ting River	145	
maximum recorded	1,518	78

Source: Kuo Ching-hui, "Silt of the Yellow River."

Table 1-12. *Origins of Annual Silt Runoff at Shan-hsien*

Area or Tributary Basin	% of Total at Shan-hsien
Above Hsun-hua	2.8
Above Lanchou	6.8
Above Pao-t'ou	10.9
Between Pao-t'ou and Lung-men	49
Wu-ting Basin	9.4
Between Lung-men and Shan-hsien	40
Wei Basin	34
Ching Basin	18
N. Lo Basin	6

Source: Kuo Ching-hui, "Silt of the Yellow River."

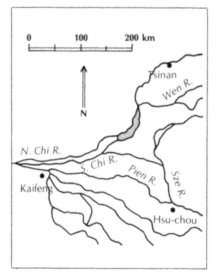

**Chou Dynasty Channels
(before 602 B.C.)**

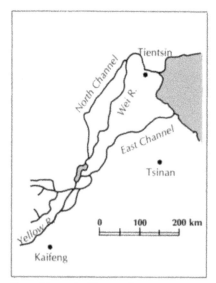

**Tang Dynasty Channels
(7th–10th centuries A.D.)**

**Sung Dynasty Channels
(after 1048 A.D.)**

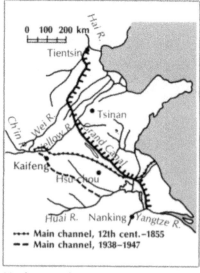

**Modern Drainage and Other
Recent Channels**

Map 1–5. *Channels on the Yellow River Floodplain in Historical Times*
Sources: Ts'en Chung-mien and HHC.

During flood crests, therefore, silt can be at least as much of a problem as high water discharge. Much of the huge silt burden is lost as the flood crest reaches the plain. Such deposition accounted for the Yellow River near Chengchou changing its channel laterally three kilometers (within the dikes) during a ten-hour period of high water in 1954.[13]

This action of the combined discharge of water and silt has been the cause of the Yellow River's rampages throughout history. Dike breaches proverbially occur "two years out of three," and in the past 3,500 years there have been 1,500 major inundations, 20 significant channel alterations, and 6 course changes so great as to move the river's mouth from one side of the Shantung Peninsula to the other.[14] Map 1-5 illustrates how radically the drainage pattern on the North China Plain has been altered by such course changes during historical times.

A permanent solution to the silt problem—controlling and decreasing the rate of erosion on the Loess Plateau—is most difficult to achieve because of climatic factors. Increasing aridity northwestward across the plateau (Map 1-3) provides for increasingly sparse natural vegetation cover. Without vegetation cover, or other means of holding the soil in place, the rate of erosion and the amount of silt in the river must remain essentially the same. Like attempts to stabilize the river's channel downstream, re-vegetation programs and other means of combatting erosion must be seen as attempts to alter the natural environment of the basin to one more favorable for human habitation.

Chapter 2
Historical Management Strategies

Early water resources management in the Yellow River basin is a subject which has attracted considerable interest among Western scholars of Chinese society and history. One debate centers on the role played by water management in shaping early Chinese civilization,[1] and another centers on the question of the origin of irrigation in north China.[2] The purpose of this chapter is not to participate in these debates, but to identify the way in which early projects attempted to solve the specific water management problems in the Yellow River basin. Information is fragmentary on the earliest projects, but there is enough information available to make possible some generalizations about the purposes, means, and agencies involved in development. Since some modern water management problems are similar to historical problems, a review of early management techniques provides valuable background for understanding modern methods.

Early Levees and Canals

The construction of levees for flood prevention apparently began much earlier than did the construction of canals for other water management benefits. Table 2-1 lists early levees identified as having been built before the sixth century B.C. Unfortunately, information is lacking on the size of these levees and the method of construction that was utilized. It is significant that most of them were located along tributary streams rather than along the main stem of the Yellow River on the lower course. Small levees were constructed to control tributary streams and were logical precursors to the larger undertakings represented by major dikes along the Yellow River itself built later in history.

Table 2-1. *Early Levees in the Yellow River Basin*

Project	River	Location (modern name)
Wen-shui Levee	Wen River	Western Shantung
Yi-shui Levee	Yi River	West of Tientsin
Chang-shui Levee	Chang River	Hopei-Honan border
Fu-shui Levee	Fu River	Southwest Hopei
Lo-shui Levee	Northern Lo River	Southeast Shensi
P'ing-yin Levee	Yellow River	Southwest of Tsinan
Cheng Levee	Yellow River	West of Chengchou
Hua-yin Levee	Yellow River	Southeast Shensi
Shang-ch'iu Levee	Chi River	Southeast of Kaifeng
Fen-shui Levee	Fen River	Southwest of Taiyuan

Source: Yang K'uan, "Achievements of Hydraulic Engineering in the Warring States Period (Chan-kuo Shih-tai Shui-li Kung-ch'eng ti Ch'eng-chiu)," in Li Kuang-pi and Ch'ien Chun-yeh, eds. *Essays on Chinese Scientific Discoveries and Scientists (Chung-kuo K'e-hsueh Chi-shu Jen-wu Lun-chi).*

These levees represented single purpose development, although their function eventually went considerably beyond the original purpose of keeping flood waters off the lands of those who had built them. During the late Chou Dynasty they came to be used as implements of war between various low-lying feudal states. The history of the Warring States period (404–221 B.C.) includes many instances of one state using dikes to divert flood waters and bring ruin on its neighbor. The fact that a treaty banning such misuse of levees was signed in 651 B.C. shows that the practice was well developed by that time, and suggests that the original use of levees to prevent floods from entering fields was probably much earlier.[3] Eventually all of the states threatened by such attacks built protective dikes. By the fifth century B.C., dike construction techniques had been extended beyond their original purpose to include the construction of protective walls between kingdoms. The Great Wall between China and the nomads on the north is entirely removed from the riverside origins of these techniques, but it represents their culmination in public works construction.

The descriptions of early canals are more detailed than those of levees (Table 2-2), but the information on canal projects is not complete. The earliest significant canal project is the Cheng State Canal, completed in 560 B.C. As with most of the other early irrigation projects, the area of irrigated land is not recorded. All that is known is that the canal watered an area somewhere in the north central portion of modern Honan. Ho p'ing-ti believes that its small scale and the difficulties associated with its construction are indicative of a lack of experience with canal building. He concludes that it was one of the first significant irrigation projects.[4]

The Han Kou Canal, built in 486 B.C. by the state of Wu to link the Yangtze with the Huai River, was extended in 482 B.C. into the Yellow River watershed. The route first headed north, near the present Grand Canal, then turned west to the Chi River, which was the name of the main branch of the Huang Ho at that time.

A century later, the Hung Kou Canal was built roughly parallel to and north of the Han Kou Canal. In 362 B.C. the kingdom of Wei moved its capital to the site of modern Kaifeng and began digging the Hung Kou Canal the next year, taking nearly twenty years to complete the project. The first section of the canal was used to drain water from the mainstream into a small lake south of the Yellow River. Later a second ditch was made to empty from the lake back into the river farther downstream. Eventually the Hung Kou Canal was extended to cross six of the small kingdoms and four major rivers to the southeast, and to connect with the Huai drain-

Table 2-2. *Early Canals in the Yellow River Basin*

Project	Date (B.C.)	Location (modern name)	Primary Purpose
Cheng State Canal	560	Northern Honan	Irrigation
Han Kou Canal	482	Southwest Shantung	Navigation
Hung Kou Canal	350	Eastern Honan-Western Shantung	Flood diversion
Chang-shui Canal	300	Chang River, Northern Honan	Irrigation
Cheng-kuo Canal	246	Ching River, Shensi	Irrigation

Source: Yang K'uan, "Achievements of Hydraulic Engineering."

age system. Two hundred years later, in the Han Dynasty, the canal was still operating and was called the Pien Canal.

Near the end of the fourth century B.C., another project with irrigation benefits significantly greater than those of any previous project was completed by the state of Wei in what is now northern Honan. A series of twelve canals was used to divert water from the Chang-shui, a tributary of the Wei River (on the Huang Plain). The system irrigated an area about five miles in length to the north of modern An-yang. This was an area significantly larger than any irrigated by a previous project, and it was used to improve the productivity of highly alkaline land.

The most famous of all the early projects is the Cheng-kuo Canal, completed by the state of Ch'in in 246 B.C. With an irrigated area of approximately 200,000 acres,[5] it was the greatest irrigation project of its time in China. Water from the Ching River was diverted to irrigate an area on the left bank of the Wei River, and eventually as far east as the Northern Lo River. The project led to vast increases in agricultural production, which contributed directly to the Ch'in state's eventual supremacy over the various warring kingdoms. The new wealth was the economic foundation of the advanced Ch'in military force that conquered all rivals and brought about the unification of the Chinese Empire under Ch'in Shih-huang-ti.

Like the levees, these canals represented purely local attempts to solve local water management problems. They primarily diverted tributary streams for single purpose development, although canals may have been used for both irrigation and navigation, as was the case later in history. The Cheng-kuo Canal, largest of the early projects, was developed on a scale that would be significant today. Even this project, however, did not represent water resources development of importance beyond the local area that it served. This irrigation scheme increased agricultural production dramatically in a portion of the Wei valley north of Sian. It did not solve the larger problems of flood and silt in the entire river system. Indeed, within several score years, the Cheng-kuo Canal itself was badly silted.

After the initial unification of China, the ruling house of the Ch'in Dynasty rapidly lost its grip on the empire it had founded. During the Han Dynasty which followed, however, the consolidation of social and political forces continued, and the Han (B.C. 206–220 A.D.) was the period when imperial government took firm roots in China. In water management, as in other important endeavors, patterns were established during the Han period that were continued, albeit with significant modifications and interruptions, throughout the entire imperial period until the first decade of the present century.

Some forty major water projects were completed on the Yellow River and its tributaries during the Han Dynasty, compared with less than ten projects in pre-Han times.[6] Many of these projects were new irrigation systems in various parts of the middle and lower basin. Southwest Shansi, the Honan-Hopei border, and the smaller streams of western Shantung all were areas where new projects developed to increase agricultural production. Transport canals were a second fact of Han development. About 130 B.C., for example, a canal was built which cut two-thirds of the time required for shipment of grain from the northern part of the Huang Plain to Changan in the Wei Valley. Such facilities were important in connecting the Han capital to the floodplain where the most rapid agricultural development was taking place.

Expansion of irrigation works toward the northwest was a third facet of Han development. In four separate prefectures of modern Ninghsia and the Great Bend area, agricultural reclamation was carried out on the basis of irrigation works on the Yellow River.[7] This initial foray, away from the low-lying valleys where Chinese agriculture had been established for several centuries, proved to be impermanent. But throughout the imperial period strong dynasties followed suit, pushing their economic frontiers to this arid fringe of the Gobi Desert.

Renovation of existing water works also became important during the Han Dynasty. The Cheng-kuo Canal irrigation system had become so silted within a century and a half as to be virtually useless. Supplementary canals were dug in 111 B.C. and a new main canal was cut in 95 B.C. Such efforts foreshadowed the way in which increasing amounts of water management energy were to be expended during ensuing dynasties. The heavy silt load of the rivers in the basin cut the effectiveness of transport, flood control, and irrigation works alike. In succeeding dynasties, new projects were opened up in various provinces of central and south China, but in the Yellow River basin the challenge had become one of redoing what the river system very quickly had undone. It had become a constant struggle to maintain an equilibrium with the river.

Nowhere was this struggle more difficult than in flood control

work. Flood control on the great plain took on increasing importance during the Han Dynasty. The greater amounts of human effort
it required were indicative of future water management problems
within the basin. Settlement on the Huang Plain had become both
more widespread and more dense during the Han Dynasty. Floods
and course changes affected more people and caused more damage than ever before. It is not surprising, therefore, that documents
of that period show increasing emphasis on the construction of
levees.[8] A memorial submitted to the Emperor Ai Ti (6–1 B.C.)
states that in a distance of about thirty-three miles near the modern Honan-Shantung border, the river was turned westward twice
and eastward three times by dikes constructed to regulate its flow.
The memorial describes how overembankment at that early date
had already begun to complicate the flood problem.[9]

The scale of water management projects was larger during the
Han Dynasty than it had been previously; larger levees were required to control the Yellow River itself than to control tributary
streams, and larger irrigation projects were required in places such
as Ninghsia and Ho-t'ao. The purposes of development were the
same as they had been prior to the Han period, however, and they
remained fundamentally the same until well into the present century. Flood control was the basic task on the lower course; without
flood control other types of development were meaningless. Navigation was an important benefit when the condition of levees was
properly maintained, but navigation canals waxed and waned with
the degree of repair or disrepair of the levee system.[10] In the middle portion of the basin irrigation was the first priority, for without
irrigation the basic productivity of the land was in jeopardy. As for
the methods of attaining these purposes, and the agencies responsible for development, patterns also were set prior to Han times
which carried through the imperial period. The methods can be
characterized as labor-intensive construction, and the agencies can
be characterized as a combination of local and national units.

Labor-Intensive Construction

The mass mobilization of laborers for earth-moving projects is the
aspect of Chinese water management that has most intrigued Westerners. In the West, the history of construction engineering has

largely been a chronicle of technological development, but in China the basic machine for moving earth did not change for nearly two thousand years. The same types of wheelbarrows first used in the Han Dynasty may be seen in photographs of dike construction on the North China Plain taken in the last few decades.[11] While technology remained relatively constant in China, progress occurred, or at least success was measured, in terms of the efficiency of labor organization.

Information is difficult to obtain on the number of workers involved in the earliest Yellow River projects. There are references to "several tens of thousands" of laborers being mobilized for individual projects that required two or three years to complete in the second century B.C. By the seventh century A.D., five million men and women were mobilized over a period of twenty years to construct one canal near the site of the earlier Hung Kou Canal.[12]

The social and political institutions utilized for labor organization have been of greater interest to Western scholars than the size of labor forces. Much attention has been focused on the process of centralization which resulted in the unification of the empire, and which was related to water management projects of increasing size. Disagreement arises over which social forces were the causes and which were the effects of the process,[13] but, contending interpretations aside, it is obvious that with unification of the empire, larger projects were undertaken by labor forces much larger than those that could be mobilized under the earlier feudal kingdoms.

The dynamics of labor mobilization also present interesting questions. Corvee labor, contributed by peasants in the agricultural slack seasons, was organized and directed by the government. The question inevitably is asked: were the laborers forced to take part in projects, or merely encouraged and helped by the government to do what they already knew to be in their own best interests? An objective answer may be that both extremes, and many situations between the two, occurred at different times in Chinese history. At times rapacious officials exploited the people,[14] and at times positive mobilization techniques were used, in which rewards were given to the most productive units and from which the people obviously benefited.[15] We may infer that the relationship between a given government and the people was critical in determining the atmosphere under which workers were mobilized for construction projects.

River Management Agencies

The study of the agencies responsible for planning, building and operating water management projects is very advantageous in understanding the overall river control effort. As the units responsible for development, these agencies represent the link between society as a whole and the resource base. During the Chou Dynasty the Ministry of Public Works (Ssu-k'ung) appears to have been the agency responsible for water management projects. Hsun Tzu (third century B.C.)[16] enumerates the duties of this office: repairing levees and bridges, clearing irrigation canals, draining away flood waters, taking care of water storage, and insuring that the people had cultivable land in times of dike breaches or droughts.

Needham mentions other officials of Chou times who functioned as inspectors of rivers and canals, or as police who patrolled embankments.[17] Beyond these fragmentary descriptions we know very little about the functions of water management agencies in pre-Han times. From the scale of the projects undertaken, and from the territorial extent of the small kingdoms of the time, we may safely conclude that the scope of such agencies' work was roughly comparable to that of modern county administrations.

With the Han consolidation of power in a central government for all the empire, a new office was created which was called Director of Water Conservancy (Tu-shui). As water management gained increased importance through subsequent dynasties, this office became more powerful and autonomous, although it remained under the authority of the central Ministry of Public Works. It is important to emphasize that this agency functioned largely as a planning and controlling body for all water management efforts in the various river basins of the empire. Labor mobilization for construction on individual projects was the responsibility of local agencies.

It is in the context of this central water management agency that the famous Yellow River managers of Chinese history appear. The famous managers were key officials who emerged at specific times and made outstanding contributions to the management of the Yellow River. The earliest of such men, from pre-Han times, were Pai Kuei of the state of Wei, remembered as the outstanding strategist at using dikes for warfare, and Cheng Kuo, the famous minister whose name was given to the project that brought so much success to the state of Ch'in.

During the Han Dynasty two men were especially prominent in Yellow River management affairs. The first of these was Chia Jang (fl. 6–2 B.C.). During his career three serious breaches of the lower Yellow River levees occurred within about forty years, and the river's mouth shifted from the vicinity of modern Tientsin to near the present mouth.

Chia Jang advocated returning the river to the abandoned course, but that channel was so seriously silted that it proved impossible. His secondary strategy was to dissipate the power of the river by draining off water for irrigation at various places in Shansi and Hopei. This was to be supplemented by diverting relatively silt-free streams into the Yellow River to increase the silt-carrying capacity. As a third measure, Chia Jang advocated strengthening dikes to contain the river. Although his primary proposals did not meet with success, Chia Jang is remembered as the first great strategist against the river, and his plans were discussed by generations of river managers after him.[18]

Wang Ching (fl. 58–76 A.D.) is credited with stabilizing the channel of the Huang Ho so effectively that there were no major dike breaches for nearly one thousand years. His major methods included dredging, strengthening the levees at dangerous points, digging new channels for tributaries in rough terrain, and building numerous sluice gates. Some writers have criticized him for not attempting to restore the river to the channel that emptied at Tientsin, but others have praised Wang Ching as one of the most efficient planners of all the famous managers.

During the Yuan Dynasty (1271–1386 A.D.) an official named Chia Lu brought order to a chaotic situation in the Yellow River basin. By the early eleventh century A.D., conditions on the lower course of the river had deteriorated to the point that dike breaches once again became frequent. In 1194 a serious flood resulted in part of the flow going south to drain through the Huai River into the sea, while the other part flowed to the north of the Shantung Peninsula. The management strategy used by Chia Lu was to stabilize the channel flowing to the Huai and to strengthen the northern dikes. His wisdom is said to have surpassed that of most managers, but he lived at a time when political disorders prevented the exercising of his full capabilities. After his career, the entire flow of the river reverted to a northern course.

During the Ming Dynasty (1368–1644 A.D.), P'an Chi-sun (1521–1595 A.D.), who has been considered by many the most able manager of the Yellow River, advocated building strong dikes to contain the river so that it would scour its own narrow channel. This was the first forceful statement against the ancient principle of dividing the flow to dissipate the river's power. P'an stabilized the main channel, which was flowing southeast at the time, and brought order to the chaotic tributaries of the Huai River.

In late Ming times, the channel once again fell into disrepair. In the early Ch'ing Dynasty (1644–1922 A.D.), Chin Fu, following the strategy and philosophy of P'an Chi-sun, stabilized the southeast-flowing river. In 1855, however, a major course change took place, and the river entered the channel which it now follows to the Po Hai.

It is important to point out the combination of water management strategy and philosophy embodied in the work of the famous managers. The question of flood control strategy focused on whether to confine the Yellow River between high levees narrowly spaced (one to three kilometers apart), or whether to give the river a broad channel between widely spaced lower levees (five to ten kilometers apart). The narrow channel accelerated bank erosion. Wide levees were able to accommodate greater flood levels, but they posed a more serious silt accumulation problem. Furthermore, peasants moving in to cultivate the rich silted land between wide levees were in constant jeopardy from high flood waters.

These two strategies represent more than different technical approaches to controlling the river. Their roots lie in different philosophical outlooks. Needham associates the construction of close, strong dikes with a Confucianist tendency to curb nature, analogous to the reliance by this school of thought on strict ethical codes for shaping human behavior. He associates widely separated, low levees with the Taoist approach of letting nature follow its own course.[19] This interpretation places water management practices squarely within the context of the major contending schools of thought in Chinese history. Chia Jang represents the "Taoist" school, while Wang Ching and P'an Chi-sun were famous representatives of the "Confucianist" approach.

Though they differed in this way, nevertheless both schools of water management gave primary allegiance to ancient precepts such as "controlling the river in accordance with the nature of water" and "using the river to control the river."[20] Each school held that it was following the correct interpretation of these precepts, which predate both Confucianism and Taoism and have their roots in the legends surrounding the first manager of Chinese waters, Yu the Great. Although it requires no further discussion here, this underlying unity of water management philosophy will provide a valuable reference point for interpreting the philosophy reflected in modern Yellow River management.

Organizationally the famous managers represented the involvement of the central government in Yellow River management. The central government had considerable interest in successful control of the great river, of course, and following the early examples of the emperors Ch'in Shih-huang-ti (221–206 B.C.) and Han Wu-ti (140–86 B.C.) many rulers took a personal interest in the management efforts. Institutionally, however, neither the managers nor their offices held basic responsibility for Yellow River development construction. These officials were inspectors of dike works and canals, and the agencies were responsible for overall planning. Labor mobilization and other facets of construction were the responsibility of provincial or lower level agencies. The River Affairs Bureaus (Ho-wu-chu) which retain responsibility for dike maintenance in the modern provinces of Honan and Shan-tung, are extensions of the provincial agencies historically responsible for development construction.

Successful management of the Yellow River, therefore, required coordination between national offices for planning and inspection and local offices for construction and operation. This was the basis of the cyclical struggle of the Chinese people with "China's Sorrow." The river could flood at any time if extreme precipitation or silt accumulation conditions existed, but the possibility of disasters occurring was considerably diminished if the dikes were in good repair and the channel adequately dredged. Efficiency in the central government's efforts to coordinate levee maintenance operations kept the river within bounds and limited the damage it could do. Corruption or inefficiency within the government led to deterioration of the dike works and the devastation of millions of acres and homes. In periods of extreme political dislocation management officials purposely let levees fall to ruin so they would receive larger appropriations, which were then squandered and peculated.[21]

The problem of coordinating river control works was made more difficult by the fact that construction was vested in provincial offices. The primary concern of local officials was that the river not flood their lands. As long as this was accomplished, they often did not care if their dike works contributed directly to disasters farther downstream, or even on the bank opposite them.

The famous manager, therefore, was more than a superior engineer. The technology of management and the basic strategy options were known to all the managers. Success required the formulation of a good plan and, more important, the ability to implement the plan. The successful manager was one who was able to mobilize and coordinate the human and material resources of the entire floodplain in a single concerted effort. In view of the difficulty of attaining that goal, it is not surprising that very few men succeeded during a period of two thousand years.

A question arises as to the necessity of the existence of a water management agency at the basin level of the organizational hierarchy. Responsible units existed at the central and provincial levels, but what of an intermediate unit with basin-wide jurisdiction, such as those which have been developed in modern times?

Nothing approximating such a unit of organization evolved until relatively late in history. During the Ming and Ch'ing dynasties the Director General of Waterways (Ho-tao Tsung-tu) was an office with responsibilities for management of the Grand Canal and the lower Yellow River basin. Its duties to a large degree lay in coordinating antiflood work among the provinces of the lower Yellow River, but the purpose of this action was to prevent damage to the Grand Canal, which crossed the Yellow River in Shantung Province. This agency is described as no more than an adjunct of the Grain Transport Administration, which relied upon the Grand Canal for northward movement of the grain tax.[22] There are also other reasons why it cannot be considered a basin administration agency: it was concerned only with flood control and not with other types of development; it had no jurisdiction over the middle and upper portions of the basin which, of course, would be necessary if an integrated basin-wide management effort were to be mounted.

This agency, furthermore, illustrates the disadvantages of a basin-level agency better than it illustrates the advantages. The Ho-tao Tsung-tu was quite successful in the Ch'ing Dynasty, when it acted only to coordinate the construction efforts of provincial bureaus. Through a process of bureaucratic growth, however, the staff of this agency was multiplied by more than a factor of ten, and it acquired its own enormous labor force for river construction work. Much of the blame for the disastrous state of Yellow River control in the

late Ch'ing Dynasty is given to precisely this process of overbu-
reaucratization.[23] The usurpation of provincial construction responsibilities by an agency less vitally concerned than local governments
with successful river control was a large part of the problem. This
question of whether certain water management responsibilities are
to be assumed at the provincial, basin, or central government level
is another one which has significance for the modern management
scheme.

Chapter 3
Early Western Interest in the Yellow River Problem

Contact between China and the West during the decades following the Opium War (1839–1842) initiated the technical and economic revolution which is still going on in China today. Although the impact of Western technology affected Chinese water management somewhat later than it did communications, manufacturing, and military development,[1] the stages of contact were roughly identical. After an initial period in which foreign water specialists visited China, Chinese students began to go abroad to acquire technical training in hydraulic engineering. Upon their return, these students worked for a time with foreign agencies in China. Through the years of internal turmoil and the war with Japan, and during the decade of close cooperation with the Soviet Union, Chinese expertise continued to develop, with training being accomplished increasingly inside China.

The withdrawal of economic and technical assistance by the Soviet Union in 1960 terminated nearly a century of foreign involvement in Yellow River management. During this century, numerous developments occurred in the collection of scientific hydrological data, in hydraulic engineering, and in water management organization. By the time of the Soviet withdrawal, therefore, a strong basis for modern development had been established, and the Chinese have proceeded with the modern plan for Yellow River management.

The Introduction of Western Science

The problems of managing the Yellow River did not attract much attention among foreigners in China until the latter half of the nineteenth century because the colonial powers were concentrating on mercantile activities; thus the disruption of agricultural life and production caused by disastrous floods went relatively unnoticed. Beginning in the 1860s, however, travelers' accounts describing the great river began to appear in Western journals. Of particular interest was the major Yellow River course change which had occurred in the 1850s. These articles varied as widely in the quality of their information as did the technical expertise of their authors. Of greatest use to later engineers were the descriptions of the river by G. S. Morrison, an English engineer, and the Dutch engineers Van Shermbeck and Visser.[2]

During the early decades of the twentieth century, foreign scientific and educational communities became more interested in the problems of China, and more attention was given to the Yellow River. The problem of its management eventually was studied by some of the most prominent Western hydraulic engineers. In 1917 and 1920 John R. Freeman, who later became president of the American Society of Civil Engineers, visited China at the request of the Chinese government. Freeman presented certain views on how flood control dikes on the Yellow River floodplain could be built most effectively.[3] These views touched off a debate between himself and two German engineers, Hubert Engels and Otto Franzius, on the question of whether the dikes should be constructed close together or far apart. Freeman and Franzius advocated building the levees closer together, while Engels favored building them farther apart to make a wider channel. The Germans ultimately used hydraulic model test experiments to support their contentions, and the views of several other prominent engineers were added to the debate.[4]

It should be noted that this discussion dealt with the same question which had occupied managers of the Yellow River since the time of Chia Jang in the Han Dynasty—the question of levee construction for flood control on the lower course of the river. Scientific study of the levees did perhaps assure a more efficient flood control system, but it also kept attention focused on only part of the entire Yellow River problem. Foreigners recognized very early, as the Chinese had for centuries, that silt was the root cause of most Yellow River difficulties,[5] yet they did not seriously discuss basin-wide solutions in the first two decades of this century.

A number of foreign engineers advised the Chinese government on specific Yellow River projects beginning early in this century. H. Van der Veen was a consultant to the Chinese Interior Ministry on solutions to the devastation caused by the 1917 flood. Captain Tyler of the British army, who spent more than twenty years in China, was instrumental in establishing the Chih-li (Hopei) River Improvement Commission. One of the most eminent of engineers was O. J. Todd, who first worked for the China International Famine Relief Commission in the early 1920s. Along with Siguard Eliassen, Todd was a consultant to the Chinese government in the mid-1930s, and he worked on closing the Hua-yuan-k'ou dike breach in 1946–1947.[6] The League of Nations sent technical teams to China in the 1930s which made various reports but took no action.[7] E. W. Lane focused primarily on methods of controlling the Huai River, but later became interested in management of the Yellow River. W. C. Lowdermilk and James Thorp, specialists in forestry and soil science, respectively, were consulted on conservation plans for the middle part of the basin.[8]

As foreign contacts became more numerous and Chinese expertise in hydraulic engineering increased, more consideration was given to a basin-wide scheme for solution to all of the Yellow River management problems, based on erosion prevention on the Loess Plateau. Few concrete measures could be implemented during the war with Japan, however. After the war, several American engineers were involved in developing preliminary plans for a basin-wide scheme. Eugene S. Reybold, James P. Growden, and John L. Savage made up a Yellow River Consulting Board that assisted in carrying out surveys and drawing up initial plans.[9] John S. Cotton, as advisor to the Chinese Public Works Commission, prepared a preliminary report on the project.[10] During the war one section of the East Asian Research Institute, organized by the Japanese occupation forces, also gathered data and formulated plans for development of the Yellow River.[11]

While foreign engineers were consulting on Yellow River problems, the number of Chinese hydraulic engineers who were assuming more of the basic water management responsibility was growing. Only a small percentage of the Chinese students who went abroad in the late nineteenth and early twentieth centuries had studied engineering. By 1935, however, twenty-seven Chinese universities and technical schools had engineering departments. A few of these, such as the Nanking School of Riverine and Oceanographic Engineering, and the Honan Hydraulic Engineering Technical School, specialized in training water management specialists.[12] At the same time, water resources research and planning bureaus were being established by governments of the Yellow River basin provinces and by the central government.

The application of modern hydraulic engineering to Yellow River basin problems was shaped by some of the most prominent engineers in China. Foremost among these were Li Yi-chih, Shen Yi, and Chang Han-ying, whose careers reflect their dedication to solving Yellow River problems. Li Yi-chih (Li Hsieh) was born in Shensi in 1883.[13] He went to Berlin in 1909 to study civil engineering. After a brief return to China, he went back to Germany in 1913 for specialized study in hydraulic engineering. Three years later he began teaching at the Nanking School of Riverine and Oceanographic Engineering. Li remained involved in education throughout his career as an engineer, ultimately holding a professorship at Peking University and the presidency at Northwest University. Equally important, however, were his accomplishments in applied engineering. Li was instrumental in formulating projects to manage the Hwai and the Yangtze, as well as developing projects on the Wei, Ching, and Northern Lo tributaries of the Yellow River.

Shen Yi studied engineering with Professor Engels in Dresden, returning to China in 1926. He held various administrative positions, including the chairmanship of the Public Works Commission at the time when the American engineers were advisors to the Yellow River project after the war. Shen went to Taiwan with the Nationalist government in 1949, where he published a collection of articles debating the merits of the floodplain dike system.[14]

Chang Han-ying received a master's degree in civil engineering from Cornell University in 1925. After returning to China he became interested in the problem of the Yellow River, even while working on other river management projects. He was chairman of the Yellow River Water Conservancy Commission (YRWCC) in 1941, and he held other administrative posts before being appointed Vice-Minister of Water Conservancy in 1950. In these various positions he has played an important role in planning and implementing the modern scheme for long-range, multiple purpose management of the river.

Of these three engineers, it was Li Yi-chih who was most instrumental in changing Yellow River management methods from traditional to modern. His writings clearly reflect the issues involved in this transformation.

Li was the first engineer, for example, to forcefully advocate an integrated basin-wide solution of Yellow River problems. While the foreign engineers were discussing the technical aspects of different dike arrangements on the floodplain in the early 1920s, Li urged that attention be focused upstream to attack the silt problem at its source. In several different articles he outlined a plan for survey, research, and construction efforts throughout the basin for integrated, multiple purpose development.[15]

Li also called for the creation of a single management agency to be responsible for planning and implementing a Yellow River development scheme.[16] Such an agency would unify administration and expenditures at the basin level. Being directly under the central government, it would be free of entanglements and squabbles between provincial governments. When the Yellow River Water Conservancy Commission was formed in 1933, Li Yi-chih was appointed concurrently director and chief engineer. This commission reflected a solution in concept only, however, because it was never able to execute the work which Li and others considered necessary. Li died in 1938 with many accomplishments to his credit, but a realization of the desire for unified basin administration was not among them. Even in the years 1945–1949, the YRWCC, with all the accumulated expertise and experience at its command, was not given full responsibility for developing the plans. Special commissions and advisory boards were appointed by the Supreme Economic Council of the Nationalist government to perform this function.

The philosophical context of modern water management reflected in Li's writings is also noteworthy. While insisting that scientific methods of surveying and data collection be strictly followed, Li observed that this was a modern application of ancient principles. Hydrological characteristics of the river system should be found in precise measurements of the "Nature of Water," the understanding of which was fundamental to management strategies. With a comprehensive knowledge of these characteristics, one is in a better position to manage the river in accordance with the Way (Tao) of water, and to apply the Great Yu's principle of "noninterference" with nature.[17]

The vision and accomplishments of Li Yi-chih certainly would rank him among the famous managers of the Yellow River, and he had other attributes that were equally important. His writings demonstrate that he was a concerned humanitarian and patriot as well as an outstanding engineer. Two of his articles, "Social Questions for Engineers" and "The Several Qualities of an Engineer," discuss how engineering projects, whether water works, highways, or railroads, must be utilized to enrich the country and benefit the people, rather than merely to serve the narrow interests of a few entrepreneurs.[18]

The Background for Basin-Wide Development

The three decades from 1920 to 1950 were an important formative period for overall management of the Yellow River. Because of political turmoil, war with Japan, and civil war, few specific projects were completed in this period. The growing corps of Chinese engineers, with the assistance of foreign advisors, did lay some groundwork for future efforts, however. A degree of progress was made in surveying, soil and water conservation research, and preliminary planning for major dam construction in the middle portion of the basin. The only concrete accomplishments of this period were in irrigation and navigation works in the middle basin.

Scattered local surveys were taken over parts of the Yellow River basin by various agencies prior to the establishment of the YRWCC in 1933.[19] The initial work of this commission included a topographic survey from Meng-chin to Li-chin on the lower course, and preliminary reservoir site surveys at Pao-chi Gorge, Lung-men Gorge, and the gorges between Shan-hsien and Meng-chin. In 1937 an aerial survey was conducted on the section of the river that forms the Shansi-Shensi border. During the war, surveys by the Chinese were carried out in the Huang-shui and Wei valleys, as well as in the Great Bend and Ninghsia Irrigation Districts.

These surveys, of course, were only the beginnings of the kind of work that was required for basin-wide development. A great number of military maps had been completed by both the Chinese and the Japanese during the war, but they were not accurate enough to be of use to engineers. Recommendations in 1946 called for the establishment of thirty-nine new gauging stations and the improvement of seventeen existing ones for climatological and meteorological data collection. Such basic data were not available for many key locations in the basin.

A beginning was also made in soil and water conservation research in the decades prior to 1949.[20] Although it had long been realized that no permanent solution to the Yellow River problem was possible without silt control on the Loess Plateau, concrete measures to effect this solution were not easily realized. In 1937 the Upstream Engineering Office of the YRWCC, in cooperation with the Shensi provincial government and the Northwest Agricultural College at Sian, began to take initial steps toward solving the silt problem. Visits by Thorp and Lowdermilk in the early 1940s helped to shape the soil conservation work, but it remained limited in scope and effectiveness for several reasons: the nature of the problem was complex; combining experimental work with extension services and education of the peasants was difficult in itself; the Nationalist and Communist governments both established experimental stations and antierosion projects, duplicating efforts on different sections of the Loess Plateau; and the resources available for soil and water conservation simply were not adequate. The amount appropriated by the Nationalist government for one of the smallest irrigation schemes (the Lao-hui Canal) in 1944, for example, was 45.5 million *yuan* (14 million dollars), while the total for all soil and water conservation work in the middle basin in 1945 was 17.2 million *yuan*. In spite of the problems, however, conservation work before 1949 was an important beginning, because for the first time attention was focused, and research begun, on the erosion problem.

Three soil and water conservation districts were established by the Nationalist government, under which the work was organized. The most active of these was the Kuan-chung District in the central Wei valley. Ten different stations carried out experiments on silt retention reservoirs, slope stabilization, check dams, water storage, and other conservation techniques, as well as on agricultural experimentation and extension programs.

Similar but limited work was carried out at three stations in the Lung-nan Soil and Water Conservation Experiment District in the vicinity of T'ien-shui on the upper Wei River, and at one station in the Lung-tung District on the upper Ching River in easternmost Kansu. A report from the Lung-nan District indicates the difficulties encountered in soil conservation work at the time. It describes opposition to the programs based on the peasants' belief that the government would victimize them. Because land was scarce, furthermore, peasants did not like to relinquish it for the experiment stations. The organization of the agency was also criticized for lack of good programs and strong leadership. Beyond these problems, the difficulties of procuring supplies and equipment during the war reduced effectiveness to a very low level.

Forestry work in the middle basin also was undertaken jointly by the YRWCC, the Shensi government, and the Northwest Agricultural College. It too was concentrated in the Wei Valley. Large areas of the Tsinling Range south of Sian, such as the Chung-nan Shan and T'ai-pai Shan, had been denuded of forests in historical times. What once were famous scenic forest areas had been reduced to barren, gullied slopes. Three separate forest conservation areas were planned for these mountains to "protect, manage, and renew remnants of the natural forests." The work was planned to include fire prevention, restriction of cutting, prohibition of agricultural cultivation, and general improvement of forest use practices.

In addition to these programs, more than 700,000 *mou* along the banks of the Wei River were designated as five separate areas in which trees were to be planted for river bank stabilization. Various other "National Defense Forest Areas" were used to cultivate walnut trees specifically for gun stocks during the war. At three different locations near Sian, experimental expansion of apple, peach, apricot, pear, and other fruit orchards was planned. A nursery was planned near T'ung-kuan to raise seedlings for the various forestry areas in the middle basin.

Limited in scope and hampered by various difficulties, these conservation measures were only the most meager of beginnings. Their value lay in providing opportunities to experiment with certain techniques in the conditions of the Loess Plateau, but they made no significant dent on the enormous problem of silt in the Yellow River. Appendix 2 illustrates the preliminary nature and the limited geographical extent of these early efforts. The areas of most serious erosion—the areas recognized as highest priority for soil and water conservation work—lay within the Chinese Communist area north of the Wei Valley. Agricultural programs in the area under Communist control were attempting to deal with erosion work, but these programs were not coordinated with the work described above. Although the priorities shown in Appendix 2 were published in 1946, even the most basic soil erosion survey could not be coordinated over the Loess Plateau until after 1949.

Preliminary planning for major dam sites was also given considerable attention during and after the war with Japan. Various proposals were made by different groups; each proposal was based on different priorities and called for different measures. A brief comparison of these proposals will illustrate the issues involved in plan-

ning the basin-wide management scheme (see Map 3-1 for proposed sites).

Li Yi-chih had emphasized flood control as the top priority for
overall management. Irrigation, navigation, and water power generation were given secondary priorities by Li.[21] He believed that
the most important long-range method of control was to remove
the silt from the river through conservation measures on the Loess
Plateau. Since such measures would require at least thirty years to
become effective, Li proposed that detention reservoirs be built
upstream to control the maximum flood discharge entering the
lower channel. His plan called for reservoirs on the Wei, Ching,
Northern Lo, and Fen to hold flood waters near their sources, and
he reserved judgment on the best site for a reservoir on the mainstream until more detailed studies had been completed.

The Yellow River development plans of the Japanese East Asia
Research Institute place primary emphasis on hydroelectric power
generation. Twenty-one different dam construction sites were considered between Pao-t'ou and Meng-chin under two different development alternatives. The most detailed plans were drawn up for
dams at Ch'ing-shui-ho (eighty-five meters high) and at San-men
Gorge (eighty-six meters high). Flood control was a lower priority
in this plan, although the San-men Gorge dam would have been
used to reduce maximum flow, as would two detention basins on
the floodplain. Irrigation and reclamation work would have been
concentrated on the floodplain, especially in Shantung Province.

The American engineers who were consulted on Yellow River
development after the war ranked flood control as the highest priority. The report of Reybold, Growden, and Savage emphasized the
use of storage reservoirs, however, de-emphasizing antierosion
measures in the middle basin.[22] This report, while praising the Japanese plan as a "careful, intelligent study," takes strong issue with
several of its proposals (see Appendix 3).

As an alternative to the Japanese key dam at San-men Gorge, the
three Americans proposed the Pa-li-hu-t'ung site, some fifty kilometers downstream. The reservoir formed by a dam at Pa-li-hu-t'ung
would be a gorge-type reservoir, without the large areas of level
valley bottom such as would be inundated by a reservoir at San-
men Gorge. They further recommended that large-capacity outlets
be built near the base of the proposed dam to discharge silt at appropriate times during the annual runoff cycle. These two features,

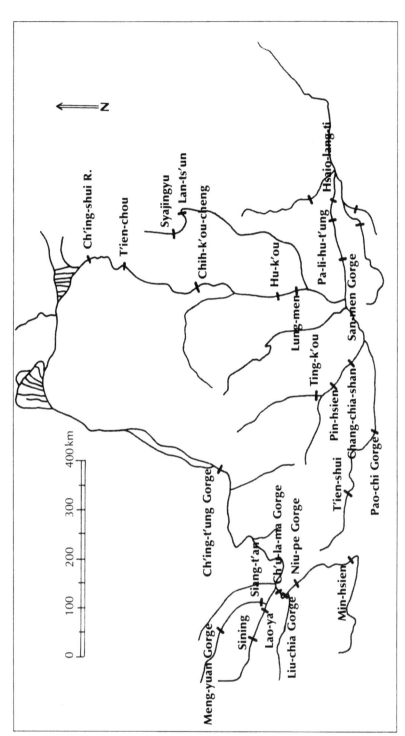

Map 3–1. *Proposed Dam Sites in the Yellow River Basin before 1950*

along with a river channel below the dam designed to accommodate discharge without deposition, would, in the Americans' view, be the most satisfactory solution to the problem of silt in the river. Similar gorge-type reservoirs with large-capacity outlets at the base of their dams were also proposed for irrigation benefits on the various tributaries. Favorable sites were known to be available elsewhere on the Yellow River as well as on the Wei, Ching, and Northern Lo rivers. The consultants expressed confidence that similar sites could be found on the Fen, Ch'in, and Southern Lo rivers.

John S. Cotton, the fourth member of the Yellow River Consulting Board, wrote a separate report to the Nationalist government's Public Works Commission.[23] Flood prevention was his primary concern, but he differed with his colleagues on the appropriate method of implementation. Cotton criticized the idea of sluicing silt through the lower channel and into the sea as wasteful of the basin's resources. His plan gave heavy emphasis to soil conservation beginning in the early phase of development, but it allowed for the fact that the river would continue to carry a heavy silt load until conservation measures were effective over the entire area—perhaps as long as fifty years after initial implementation. During the first phase, therefore, reservoirs at Lung-men, Chin-k'ou, and the mouths of the major Wei River tributaries would provide flood control while filling up with silt. The silt deposited in these reservoirs could then be cultivated, and would reduce the undercutting and caving of banks which is a major source of silt in the river. Before the reservoirs were entirely filled with silt, other regulation reservoirs would go into operation, including the key to the system—a reservoir at San-men Gorge. But Cotton emphasized that a dam at San-men Gorge should be built only after the others were completed, in order to prevent the siltation of its reservoir.

The crux of the differences between these early schemes lay in their proposals for handling the silt problem in relation to reservoirs constructed for flood control and other benefits. The striking fact is that construction of a dam at San-men Gorge during the first phase of development was proposed only in the Japanese plan. This plan was patently exploitative, concentrating on short-term hydroelectric power benefits at the expense of long-term solutions to the silt problem.

Of the other writers, Li made no proposal concerning San-men Gorge because of insufficient information, Cotton said the life of a reservoir at this site would be too short unless constructed after the river's silt load had been substantially reduced, and Reybold, Growden, and Savage held that such a reservoir would inundate too much valuable land, so should never be built. All these proposals were admitted by their authors to be based on insufficient data; all agreed that much more information would have to be collected before plans could go beyond the preliminary stage.

Irrigation in the middle part of the basin was one phase of development taken beyond the planning stage prior to 1949. Four separate irrigation areas were established within the portion of the Loess Plateau under the control of the Nationalist government. These four were: the Kuan-chung Plain, the Great Bend, Ninghsia, and the Kansu-Chinghai area.

The most significant development in irrigation during the Nationalist years was on the Kuan-chung Plain of the Wei Valley in central Shensi. Pre-existing irrigation systems in reasonable condition watered about 150,000 *mou*, but projects begun in the 1920s and developed during the 1930s and 1940s added the potential to irrigate some three million *mou* (see Table 3-1). Li Yi-chih was chief engineer for the first of these projects, the Ching-hui system, and the others bore the mark of his leadership. Of the major canal systems planned in this period, only the Lo-hui system on the Northern Lo River was not completed. Irrigation was developed to the point where dry season flow of the major streams was almost entirely utilized, so that storage reservoirs gained a higher priority than additional irrigation schemes.[24]

In the Ho-t'ao region projects to irrigate nearly four million *mou* were said to be operating or under construction in 1945, and they were located almost exclusively on the northwest corner of the Great Bend. Further work planned at the time was to increase the total irrigated area to over twenty million *mou* (Table 3-2). The primary task was dredging and renovation of the existing canals, some of which were first constructed in the early Han Dynasty. The improvement of these canals was a very difficult problem since there was no regulatory equipment at the intake gates. The canal heads were simply open gates which received water when the Yellow River was high and none when it was low. Flooding and silting of the main stream, furthermore, had shifted its current away from these head gates. Many of the canals themselves were silted up and the land abandoned. Eventually, a very small canal system on the K'un-tu-lun River, emptying into the Yellow River near Pao-t'ou, was to be expanded to irrigate an area equal to that of the Hou-t'ao

Table 3-1. *Irrigation Projects and Plans in Shensi before 1949*

Irrigation District	Irrigated Area (1,000 *mou*)	Construction (begun)	Construction (completed)	Canal Length (km)	Diversion Rate (cms)	Dam Height (m)
Ching-hui	730	1930	1935	252	16.0	9.2
Wei-hui	600	1935	1937	172	30.0	3.2
Hei-hui	160	1938	1942	56	8.6	
Mei-hui	132	1936	1938	142	8.0	
Lo-hui	500	1934	a	84	15.0	16.2
Feng-hui	230	1941	1947	48	10.0	
Lao-hui	100	1943	a	22	2.5	
Kan-hui	3	1943	1944	7	0.4	
Yao-hui	75	1940[b]				
Pa-hui		1938[b]				
P'eng-hui	170	1940[b,c]				
Ting-hui	30	1941	1943	34	7.0	
Chih-an	11		1938			

Source: WRPC No. 112; *SLHC.*
[a] Not finished in 1947 (*SLHC*).
[b] Plans announced in year indicated.
[c] Projected completion in first Five Year Plan (WRPC No. 71).

system. This plan, like many others, was not carried out before 1949.

The Min-sheng (or Salachi) canal project at the northeastern corner of the Great Bend was undertaken in 1928 at a site where major irrigation works previously had not been functional. It was begun by the Suiyuan provincial government, and then received technical and financial assistance from China International Famine Relief Commission. The project was designed to irrigate two million *mou*, but was a complete failure. Water was diverted at the intake fifteen kilometers east of Pao-t'ou in 1931, but siltation and other problems were so serious that it was not operational even for one season. After the war, the project was abandoned and no satisfactory proposals were put forth for its rehabilitation.[25]

In Ninghsia, approximately 2.6 million *mou* were said to be under irrigation in 1947, with plans for the amount to increase to 5.5 million *mou.*[26] In the largest irrigation district of Ninghsia, the Ho-hsi District lying west of the Yellow River and north of the Ch'ing-t'ung Gorge, major improvements had been made in 1943. In addition to expanding irrigation, the projects planned in 1947 included power generation, ship locks, bridges, and a movable

Table 3-2. *Irrigation Projects on the Great Bend before 1949*

Irrigation District	Canal System	Irrigated Area (1,000 *mou*)		
		1945	When Renovation 1947 Completed	Projected Completion Date
Hou-t'ao		3,605	10,000	
	Mi-shr		10	
	Feng-ch'i		10	
	Fu-hsing		225	
	Yang-chia		812[a]	
	Ho		1,121[a]	
	Yung-chi		1,044[a]	
	Yi-ho		628[a]	
San-hu-ho		50	440	1951
Min-sheng		0	2,500	1951
K'un-tu-lun-ho		10	10,000	1948
Ta-hei-ho		32	320	1948
Hung-ho		25	100	1951
Pao-pei-ho		5	20	1951
Min-li		10	400	1948
Yi-k'o-chao		65	1,000	1951
Main mtn. districts		40	190	1951
Small mtn. districts		100	200	1948

Sources: WRPC No. 134; *SLHC.*
[a] Under construction.

water level stabilization lock at the main canal intake. Drainage was given equal priority with irrigation in the reclamation plans, owing to perennial problems of high water table and alkalinity on the Ninghsia Plain.

In the Ho-tung Irrigation District east of the river in Ninghsia, the plans included canal improvements and a sediment lock at the intake to help alleviate siltation of the irrigation canals. The Chung-wei Irrigation District upstream from Ch'ing-t'ung Gorge was given lower priority than the other two districts, since the potential for development within its narrower valley was much more limited.[27]

In Kansu and eastern Chinghai, 950,000 *mou* on the Yellow River and its tributaries were under irrigation before 1949. Of this amount nearly 100,000 *mou* still utilized wooden water wheels as large as sixty meters in diameter to lift water from the streams. Another 350,000 *mou* were watered by small-scale local gravity flow systems constructed on the T'ao, the Huang-shui, and the Yellow rivers between 1938 and 1947.[28]

Irrigation in the other provinces of the middle Yellow River basin was largely in the planning stage. A scheme which historically had irrigated as much as 300,000 *mou* in the vicinity of Taiyuan, was in utter disrepair; but Nationalist plans called for it to be improved to the point where half a million *mou* would be watered. In the southwest corner of Shansi a reclamation project was envisioned that would irrigate and drain 115 square kilometers of highly alkaline land.[29]

Use of the river for navigation also received some attention during and after the war with Japan. Navigable sections of the Yellow River above Pao-t'ou, and of the tributaries, are listed in Table 3-4. Planned improvements for inland navigation are also indicated.

To summarize the state of Yellow River management in 1949, several decades of foreign contact had begun to lay the groundwork for basin-wide development, but very little construction had actually taken place. There was a growing corps of Chinese hydraulic engineers trained in modern management techniques, and some progress had been made in irrigation and navigation development.

Table 3-3. *Irrigation in Ninghsia before 1949*

Irrigation District	Irrigated Area 1947	When Improvements Completed (1,000 *mou*)
Ho-hsi	1,711	4,730
Ho-tung	429	790
Chung-wei	496	

Source: WRPC No. 59.

Table 3-4. *Navigation on the Yellow River and Tributaries, 1947*

River	Section	Length (km)	Status	Planned Improvement Depth (m)	Ship size (tons)
Yellow	Lanchou to Ninghsia	500	Plans being made		
	Ninghsia to Pao-t'ou	600	Navigable to 20-ton vessels	2	300
Wei	Pao-chi to T'ung-kuan	350	Badly silted, elevated channel	2	300
T'ao	Min-hsien to mouth	259	Navigable to small craft Completed 1942–1943 Width: 14 m Depth: 0.6 m		
Huang-shui	Hsiang-t'an Gorge to Ta-chia	66	Navigable to small craft Completed 1941–1945		

Sources: WRPC No. 59; *SLHC.*

However, flood control, silt control, power generation, and other benefits were only in the planning stage. Development on these fronts could not go forward until social and political conditions were stabilized and the country was unified.

Chapter 4
Soviet Assistance in Yellow River Development

Economic and technical assistance from the Soviet Union played an important role in all aspects of Chinese development in the decade between 1950 and 1960. In Yellow River management, Soviet advisors assisted in drawing up and implementing the "Multiple Purpose Plan for Permanently Controlling the Yellow River and Exploiting Its Water Resources," unveiled in July 1955.

The purposes, means, and agencies involved in this scheme all reflected the type of basin-wide development program which had been envisioned prior to 1949. The purposes were varied and ambitious, including the complete elimination of floods, the generation of hydroelectric power equal to ten times the national total in 1954, a sevenfold increase in irrigated acreage in the basin, and an increase of more than twenty times in the navigable length of the river. The primary means of accomplishing these goals was to be the construction of forty-six large dams on the main stem and numerous smaller dams on tributaries. The agency responsible for coordinating this development was the Yellow River Water Conservancy Commission (YRWCC).[1]

Implementation of this scheme during the 1950s has been characterized as "Soviet-style development" because of the heavy emphasis placed on the construction of large dam projects.[2] Large dams certainly commanded the spotlight during this period, but other features of the development program are equally important in understanding the overall impact of Soviet assistance. The implementation of a truly basin-wide approach can be illustrated by the early survey effort and by the plans for interbasin water transfer. The attempt to unify and strengthen basin-wide administration is illustrated by the operation of the YRWCC.

Implementation of Basin-Wide Planning

Many Chinese engineers and writers remarked during the 1950s on the important lesson to be learned from the Soviets regarding the use of science for successful water management.[3] Chinese technical personnel had been trained in the collection of scientific data and its use in design and other tasks before 1949, but few had seen the process in operation on a large scale. Survey work was the earliest and most direct application of the scientific approach. Table 4-1, which lists some sample surveys, is by no means comprehensive, but it does illustrate the kinds of work which were carried out in the Yellow River basin during the early years of the new government, as reported in the Chinese press.

The survey work was a continuation and expansion of what had begun two decades before and had proceeded slowly through the years of turmoil and war. A considerable number of trained personnel were available in 1950, but the task before them was enormous. Much of the basin had never been surveyed even superficially, and parts had been visited only rarely by the Chinese.[4] The survey work was undertaken with enthusiasm, however, and many of the personnel enlisted received their training on the job.

The YRWCC was responsible for much of the survey work. Fifteen teams from the YRWCC Survey Corps, for example, began in 1950 on a basic hydrologic, weather, and soil survey of the basin that extended through 1952. In 1953 ten more teams were organized jointly with various institutes of the Academy of Sciences for a basic physical, social, and economic survey of the Loess Plateau. Similarly, in 1955 an inspection team of 150 members from the YRWCC, the Academia Sinica, the Ministry of Forestry, and various universities launched a soil conservation survey in Shansi Province. Throughout the 1950s the YRWCC continued to function as a coordinating agency for units from various scientific, academic, and governmental (national and provincial) offices.

Certain special commissions were also set up for key tasks in the survey work. The Northwest Soil and Water Conservation Study Mission, for example, was formed of workers from the Ministries of Agriculture, Forestry, and Water Conservancy to assist with the 1953 Loess Plateau Survey mentioned above. A special commission of the Central Peoples' Government (CPG) carried out specific soil and water conservation surveys in 1953 and 1957 in the areas of most severe erosion. Specialized surveys also were carried out by separate groups, such as the Survey Department of the Hydroelectric Construction Bureau, which worked at the major power station locations.

Other permanent agencies with specialized tasks also were created, most of which were under the general coordination of the YRWCC. The Yellow River Research Group was organized in 1953 by the Ministries of Water Conservancy and Fuel Industry to analyze certain records on the Yellow River and to help coordinate the work of Soviet advisors. The Yellow River Survey Planning Board was created in early 1956 to concentrate technical questions in overall survey work and to do preliminary planning on certain tributary basin reservoirs.

In spite of the heavy emphasis given to survey work in the early years of the Communist government, the enormity of the task meant that it could not be completed in a short period of time. Much surveying remained to be done when the early irrigation and multiple purpose projects were begun. By October 1956, for example, only one-fourth of the basin had been covered by basic hydrologic survey,[5] although the number of hydrologic gauging stations had been increased from seventy-nine in 1947 to nearly three hundred, and more than four hundred rainfall gauging stations had been established. By early 1955 topographic surveys had been completed on only 21,000 square kilometers, less than 3 percent of the total basin area.[6] Survey work continued in the basin while other facets of the overall scheme gained the spotlight.

Interbasin transfer of water resources has proven to be the most ephemeral feature of the early planning efforts. Survey work was used to lay a foundation for all aspects of development, and large dam construction has remained a vital part of the overall strategy; but interbasin transfer, which received a great deal of attention in the late 1950s, apparently has not been taken beyond the planning stage.

The reason given at the time for focusing attention on interbasin transfer was the relative scarcity of water resources in the Yellow River basin. Indeed, all of north and northwest China account for 51 percent of the country's cultivable area, but have only 7 percent of the total river discharge. The Yangtze basin and south China, by contrast, include 33 percent of the cultivable area, but 76 percent of total river discharge.[7] In 1959 the head of the Soviet water conservancy delegation stated that the water resources of the Yellow River (as well as those of the Huai and Hai rivers) were nearly all

Table 4-1. *Sample List of Surveys in the Yellow River Basin after 1949*

Date	Area	Type/Work	Agency
1950–1952	Entire basin	Established more than 200 weather and hydrologic stations	YRWCC (15 different teams)
June 1952– Feb. 1953	Upper & middle reaches	Hydropower survey; surveyed reservoir areas, drilled dam sites, charted 19 gorges	a
Aug.–Dec. 1952	Headwaters	Topographic, geologic, hydrographic	YRWCC
Jan. 1953	Mouth	Hydrographic	a
Mar. 1953	Upper Wu-ting R. Upper Ching R.	Soil conservation	CPG special commission
May–July 1953	Major tribs. on Loess Plateau	Physical, social, economic	YRWCC survey corps; NW Soil and Water Study Mission
Apr.–June 1955	Middle basin	Hydrographic	Yellow River Planning Comm.
Apr.–Aug. 1955	Yi, Lo, and Chin valleys	Comprehensive	a
June–Dec. 1955	Shansi Province	Soil conservation survey (20,000 sq km)	YRWCC
Sept. 1955	NE Kansu	Reclamation project	a

Date	Location	Purpose	Agency
Oct. 1955–Oct. 1956	Middle and lower reaches	Soil survey, map: soil types and alkalinity	Acad. of Sciences
Spring–fall 1956	Shansi-Shensi border	Hydrologic, topographic, geologic; gather data for reservoir design	Yellow River Survey and Design Board
Apr.–Aug. 1956	Wu-ting R.	Comprehensive	[a]
Apr.–Oct. 1956	Ching R. Basin	Geographic and economic	[a]
July 1956	Inner Mongolian Autonomous Region	Comprehensive (technical, social, economic)	IMAR Water Conserv. Dept.
Aug. 1956–Feb. 1957	Shansi Province	Groundwater	[a]
Dec. 1956–Jan. 1957	Middle and lower reaches	Aerial survey	Ministry of Forestry
Sept. 1957	Shansi and Shensi	Water and soil conservation	CPG special commission
June 1958	Floodplain	Flood diversion	YRWCC
Dec. 1959	——	Route selection complete for planned diversion from Yangtze Basin to Yellow R.	YRWCC

Source: Various NCNA reports.
[a] Agency not specified in source.

utilized and that further development of irrigation could not proceed without major new undertakings.[8]

The idea of diverting Yangtze basin water northward to the Yellow River was discussed in the early and middle 1950s, although it was not treated in detail in the 1955 announcement of Yellow River development plans. In the spring of 1959 a full-scale survey of water resources and possible diversion routes was undertaken. The Huai and Hai basins were considered together with the Yangtze and Yellow River basins in a large scheme called "Southern Waters Transferred Northward (Nan-shui Pei-tiao)." Some fifteen possible diversion routes were identified, among which the following were considered most promising (see Map 4-1):[9]

1. The Yu-chi Canal extending from the vicinity of Yu-shu on the upper Yangtze to Chi-shih Shan on the upper Yellow River would be 1,688 kilometers long and would divert about 22 billion cubic meters of water annually. Several tunnels would be required, including one 6.5 kilometers in length. The canal would meet the Yellow River at an elevation high enough to supply water to arid regions of Chinghai and Kansu; however, the total amount transferred would not suffice to solve the water shortage problems of these arid regions. It also would reduce significantly the water available for hydropower generation in the Yangtze Valley.

2. The Weng-ting route, extending from the Weng-shui River in Northern Yunnan to the vicinity of Ting-hsi in Kansu,[10] would require five dams and fifteen tunnels in its 6,800 kilometer length. One hundred and forty-two billion cubic meters (13 percent of total Yangtze basin discharge) could be diverted annually. But the route would neither enter the Yellow River at a high enough elevation to benefit the Tsaidam basin in western Ching-hai nor include any sites for reservoirs large enough for satisfactory long-term storage. The geologic conditions in the steep gorges would make the tremendous amount of construction required difficult.

3. A San-hsia–Peking canal would link huge reservoirs at the San-hsia Gorges on the Yangtze and at Tan-chiang-k'ou on the Han River, then would follow the foot of the Fu-niu Shan to enter the Yellow River at the T'ao-hua-yu Reservoir near Chengchou. This section would be 500 kilometers long and require 560 million cubic meters of earth and stone work. Twenty-three billion cubic meters of water could be diverted annually from the Han River alone, but benefit would accrue only to the lower Yellow River and eastern Hopei via the Wei River on the North China Plain.

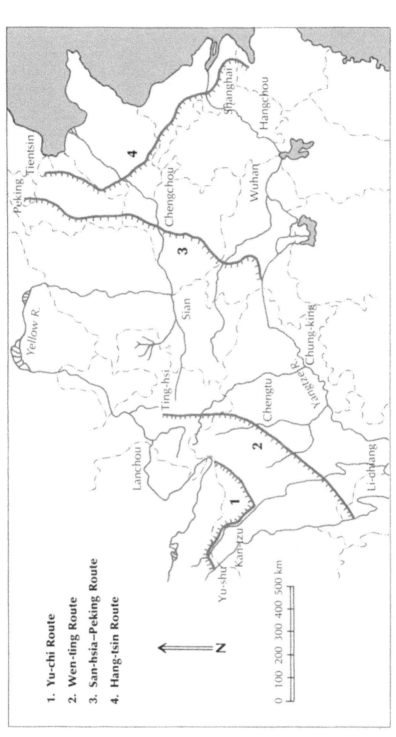

1. Yu-chi Route
2. Wen-ting Route
3. San-hsia–Peking Route
4. Hang-tsin Route

Map 4–1. *General Route Alternatives for Proposed Interbasin Water Transfer*

4. The Hang-tsin diversion route would follow the Grand Canal through various lakes near the eastern coast, linking Hang-chou and Tientsin. Ten pumping stations would be required to raise the water sixty to seventy meters. Ten billion cubic meters could be transferred annually at an average rate of thirty-two cubic meters per second. This would be the easiest of the transfers to make, but the least beneficial in terms of area served.

Most attention was focused on the second of these routes, the one linking the Yangtze River and the Han River with Ting-hsi in the Yellow River basin. Wang Hua-yun, director of the YRWCC, suggested that extremely high dams be built in the deep, narrow gorges of the upper Yangtze and Ya-lung rivers in Yunnan and Sze-chuan.[11] He described the Hu-t'iao Gorge near Li-chiang, Yunnan, as 30 to 40 meters wide with walls 5,000 to 6,000 meters high. Noting that the world's highest concrete dam at the time was 280 meters high and the highest rock fill dam was somewhat more than 100 meters high, Wang admitted that the proposal was without precedent. But he argued that earthquake-induced blockages of similar gorges produced natural reservoirs that had not leaked through subsequent tremors in more than twenty years. With water levels raised to 2,500 meters above sea level in such gorges, he pointed out, large amounts of water could be diverted to points on the Yellow River higher than Ting-hsi in western Kansu, providing the potential to solve the water shortage in Chinghai, Kansu, Ning-hsia, and Inner Mongolia.

On the question of deleterious effects to the Yangtze basin of transferring water northward, Wang argued there would be none and that certain benefits in fact would accrue, including reduction of flood crests in high water periods, shipping on the long impoundments formed in the narrow gorges, power generation and irrigation at the new dam sites, and secondary developmental impacts in the affected areas.

The completion of the enormous Tan-chiang-k'ou Reservoir on the Han River was announced in the spring of 1974. Since this reservoir was a key element in the plans for interbasin transfer the question arises as to the current status of the diversion project. Chinese officials state that no water is being transferred into the Yellow River basin and that no plans for transfer are being considered currently.[12] The reason given is that no actual water shortage exists in the Yellow River basin since other measures have been taken to alleviate this problem.

The first of these measures is more efficient use of ground water resources, and the second is storage in small- and medium-scale reservoirs throughout the basin. The development of these techniques during the 1960s will be discussed in the next chapter, but their adoption as substitutes for the grandiose transfer scheme was foreshadowed as early as 1959. It was pointed out at that time that comprehensive plans for each of the basins were incomplete, and that it would be highly impractical to pursue planning for interbasin transfer until the economic impact of such a transfer could be predicted and the technical problems solved.[13] The fact that no significant interbasin transfers have been made seems to justify that viewpoint.[14]

It is useful, nonetheless, to note the allocation of Yellow River water resources planned in conjunction with the interbasin transfer scheme, because of the view it gives of relative apportionment of water within the basin.[15]

Total annual Yellow River discharge near Chengchou is approximately 4.5 billion cubic meters. The plans contemplated 16.11 billions cubic meters of this total eventually being lost to uses above San-men Gorge, to the Yi, Southern Lo, and Ch'in basins, and to other areas between San-men and Chengchou. The 29.39 billion cubic meters remaining would be supplemented by 26 billion cubic meters annually from the Han basin, making a total of 55.39 billion cubic meters available in the lower Yellow River basin. Ultimately it would be possible, using several different canals, to transfer as much as 156 billion cubic meters annually from the Yangtze basin to the Yellow River basin.

Plans also were sketched out for the use of the 55.39 billion cubic meters in the lower basin after Han River diversion. The allocation per sector of this water is shown in Table 4-2; the provincial distribution of water resources before and after diversion is shown in Tables 4-3 and 4-4.

Emphasis on Large Construction Projects

The primary feature of the period of Soviet assistance was the construction of large-scale dams. During the first Five Year Plan (1953–1957), large, capital-intensive construction projects, often designed in the Soviet Union, characterized many aspects of Chinese economic development. In the Yellow River management scheme this

Table 4-2. *Allocation per Sector of Water Resources after
Implementation of Han–Yellow River Diversion
(Billion Cubic Meters)*

Use	Yellow River	Han River	Total
Nonproductive uses[a]	2.24	1.5	3.74
Industrial and urban	1.28		1.28
Navigation	1.39	.70	2.09
Irrigation	23.3	23.8	47.1
Fisheries	1.18		1.18
Total	29.39	26.0	55.39

Source: *Ajia no Yume.*
[a] Nonproductive water uses include:
1. 740 million cubic meters of water losses from Tung-p'ing.
2. 1 billion cubic meters of water losses from the Yellow River.
3. 500 million cubic meters of water losses from irrigation in all the provinces.
4. 1.5 billion cubic meters of water losses from diverting the courses of the Han, Chi, and Yellow Rivers.

meant emphasis on the key dams, built with state funds, which were to be the backbone of Yellow River control.

Forty-six dam sites were identified to "staircase" the main stem of the river when the plan was first unveiled (see Table 4-5 and Map 4-2).[16] Addressing the specific management problems in each portion of the basin (Chapter 1), the dams above Lanchou and in the Shansi-Shensi border region were to emphasize power generation; those in the Great Bend area and on the floodplain were planned primarily for irrigation benefits; and those in western Honan were to combine flood control and power generation. All the dams were to serve multiple purposes, however, including navigation, so that this plan of development represented a heavily capital-intensive means of reaching multiple purpose water management goals. Other, less capital-intensive projects, such as soil and water conservation and labor-intensive irrigation construction also were carried out during the 1950s (see Chapter 5), but primary emphasis was on large dams with the hope that spectacular and immediate results could be achieved.

Of the forty-seven dam sites identified, five were selected as most important for early development. The 1955 plan specified that construction at Liu-chia Gorge, Ch'ing-t'ung Gorge, San-sheng-kung, San-men Gorge, and T'ao-hua-yu was to be completed during the first phase of development, ending in 1967.[17] Construction had begun at all these sites except T'ao-hua-yu by 1958. Because of the enthusiasm generated by the Great Leap Forward, which began that same year, the construction schedule was accelerated and work reportedly began at several other sites. Included in this latter group were projects at Yen-kuo Gorge near Lanchou, and at Wei-shan, Lo-k'ou, and Wang-wang-chuang on the river's lower course in Shantung Province.[18]

Beginning in 1960, the information available on all aspects of development in China was drastically curtailed. The impact of the So-

Table 4-3. *Provincial Distribution of Yellow River Basin Water Resources prior to Han River Diversion*

Use	Hopei	Shantung	Honan	Kiangsu	Total
Irrigation					
Area[a]	47.65	80.0	80.4	8.4	216.45
Amount[b]	54.0	49.3	83.1	5.4	191.8
Municipal and industrial					
Intake rate[c]	197.6	118.5	249.0	28.0	593.1
Intake amount[b]	6.24	3.75	10.95	.88	21.83
Waste[b]	.76	2.85			3.61
Fisheries					
Intake rate[c]		50.0			50.0
Intake amount[b]		1.5			1.5
Navigation					
Intake rate[c]	130.0	285.0	150.0		565.0
Intake amount[b]	4.1	8.45	4.75		17.3
Waste[b]	.76	2.85			3.61
Other uses[b]		4.5			4.5
Total amounts[b]	65.86	73.20	98.81	6.28	244.15

Source: Ajia no Yume.
[a] Million *mou.*
[b] Billion cubic meters.
[c] Cubic meters per second.

Table 4-4. Provincial Distribution of Yellow River Basin Water Resources after Han River Diversion

Use	Hopei	Shantung	Honan	Kiangsu	Anhwei	Total
Irrigation						
Area[a]	93.1	109.0	80.4	12.8	29.3	324.6
Amount[b]	77.8	111.55	83.1	7.1	7.6	287.15
Municipal and industrial						
Intake rate[c]	197.6	118.5	249.0	60.0		625.1
Intake amount[b]	6.14	3.75	10.96	2.0		22.85
Waste[b]	3.32	2.5	2.15	0.4		8.37
Fisheries						
Intake rate[c]		50.0				50.0
Intake amount[b]		1.5				1.5
Navigation						
Intake rate[c]	130.0	285.0	150.0			565.0
Intake amount[b]	4.1	8.45	4.75			17.3
Waste[b]	0.76	7.1				7.86
Other uses[b]		13.9				13.9
Total amounts[b]	92.12	148.75	100.96	9.5	7.6	358.93

Source: Ajia no Yume.
[a] Million mou.
[b] Billion cubic meters.
[c] Cubic meters per second.

viet withdrawal of economic and technical assistance in July of that year, together with the difficulties experienced in the Chinese economy, made government authorities reluctant to release progress reports of any kind. The result was a period of about fifteen years during which almost no specific information was available on the progress of major projects in the Yellow River basin. Scattered references were made to the major dam projects, but no official announcements on their status were forthcoming.

In late 1974 and early 1975, however, a series of official announcements was made on the completion of major Yellow River dams. In September 1974, completion of the Ch'ing-t'ung Gorge project was announced.[19] The dam at Ch'ing-t'ung Gorge, above Yin-ch'uan in the Ninghsia Hui Autonomous Region, is 697 meters long and 42 meters high. The first of six power generation units was installed in 1967, and the last generator was installed in 1974, bringing the total installed capacity to 260,000 kilowatts. Irrigation trunk canals 113 kilometers long and 70 kilometers long were built

Table 4-5. *Major Multiple Purpose Dams on the Yellow River Mainstream (Planned, 1955–1960)*

Project	Dam Size (m)		Storage Capacity (billion m³)	Generating Capacity (thousand kw)
	Height	Length		
Lung-yang G.[a]	105		20	1,000
La-hsi-wa	(202)[b]		1.7	1,600
Ni-ch'iu Shan	(25)[b]		0.02	180
Sung-pa G.	(94)[b]		0.4	720
Li-kung G.	(40)[b]		0.04	300
Kung-po G.	(118)[b]		1.5	900
Chi-shih G.	140		3.41	1,000
Szu-k'ou G.	49		0.7	420
Liu-chia G.	148	1,100	4.91	1,050
Yen-kuo G.	45		0.24	595
Pa-pan G.	16		0.03	20
Ts'ai-chia G.	18		0.08	250
Wu-chin G.	74		1.64	950
Hei-shan G.	120		11.4	1,500
Ta-liu-shu				
Ch'ing-t'ung G.	42	560		260
San-tao-k'an				
San-sheng-kung				20
Chao-chun-k'an				
Hsiao-ho-wan				
Wan-chia-sai				400
Yu-k'ou				
Shih-lan				
Ch'ien-pei-hui	149	835	4.43	1,000
Lo-yu-k'ou	88	630	1.97	560
Yi-yueh				
Chi-k'ou	94	685	3.46	700
San-ch'uan-ho				
Lao-o-kuan				
Jen-yu-li				
Yen-shui-kuan				
Li-jen-p'o	97	585	2.17	720
Hu-k'ou	68	615	0.22	640
Lung-men				

(continued on next page)

Table 4-5. *(continued)*

| Project | Dam Size (m) | | Storage Capacity (billion m³) | Generating Capacity (thousand kw) |
	Height	Length		
Han-ts'un				
An-ch'ang				
San-men G.	120	963	64.7	1,100
Jen-chia-tui				
Pa-li-hu-t'ung	127	475	2.37	1,870
Hsiao-lang-ti	155	450	12.0	2,300
Hsi-hsia-wan				
Hua-yuan-chen				
Ch'u-yu				
T'ao-hua-yu	117		7.3	50
Wei-shan	15	350	4.0	38
Lo-k'ou				
Wang-wang-chuang				

Sources: NCNA; A. A. Koroliev, "The Key Yellow River San-men Gorge Water Conservancy Project (Huang-ho San-men-hsia Shui-li Shu-niu)," *CKSL*, No. 15 (March 1957), pp. 5–7; *Ajia no Yume.*
    [a] Gorge.
    [b] Indicates largest possible head for reservoir; dam height not given.

on the west and east sides of the Yellow River respectively, bringing the total irrigated area to over three million *mou*. In addition to power generation and irrigation, flood control and the control of winter ice floes are benefits provided by this project.

Completion of the Yen-kuo Gorge project, in Yung-ching County, Kansu, above Lanchou, was announced at approximately the same time.[20] The dam is 321 meters long and 57 meters high, forming a reservoir with about 220 million cubic meters storage capacity. Power generation began in 1962 at Yen-kuo Gorge, and a total of 11.8 billion kilowatt hours was generated in the first twelve years of operation, although no total installed capacity was announced. The project was built primarily to supply power to the Lanchou region, but it is also capable of irrigating 45,000 *mou* of farmland.

The giant Liu-chia Gorge project, located 25 kilometers upstream from Yen-kuo Gorge, was officially completed on February 5, 1975.[21] It is the largest hydroelectric power station in China, with an installed capacity of 1.2 million kilowatts and the country's first

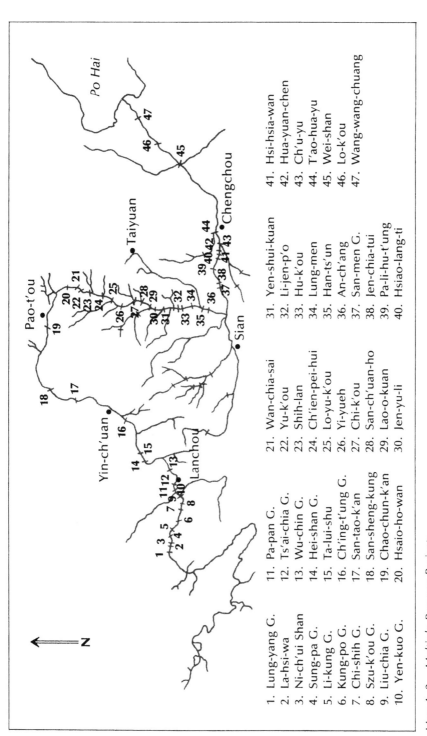

Map 4–2. *Multiple Purpose Projects*

1. Lung-yang G.
2. La-hsi-wa
3. Ni-ch'ui Shan
4. Sung-pa G.
5. Li-kung G.
6. Kung-po G.
7. Chi-shih G.
8. Szu-k'ou G.
9. Liu-chia G.
10. Yen-kuo G.

11. Pa-pan G.
12. Ts'ai-chia G.
13. Wu-chin G.
14. Hei-shan G.
15. Ta-lui-shu
16. Ch'ing-t'ung G.
17. San-tao-k'an
18. San-sheng-kung
19. Chao-chun-k'an
20. Hsaio-ho-wan

21. Wan-chia-sai
22. Yu-k'ou
23. Shih-lan
24. Ch'ien-pei-hui
25. Lo-yu-k'ou
26. Yi-yueh
27. Chi-k'ou
28. San-ch'uan-ho
29. Lao-o-kuan
30. Jen-yu-li

31. Yen-shui-kuan
32. Li-jen-p'o
33. Hu-k'ou
34. Lung-men
35. Han-ts'un
36. An-ch'ang
37. San-men G.
38. Jen-chia-tui
39. Pa-li-hu-t'ung
40. Hsiao-lang-ti

41. Hsi-hsia-wan
42. Hua-yuan-chen
43. Ch'u-yu
44. T'ao-hua-yu
45. Wei-shan
46. Lo-k'ou
47. Wang-wang-chuang

300,000-kilowatt double water internally cooled generator. The annual output is 5.7 billion kilowatt hours. As at Yen-kuo Gorge, the generation equipment was designed and manufactured in China. The first generator began operation in 1969, and the last one was installed in December 1974. Power is distributed from Liu-chia Gorge to eastern Kansu, eastern Chinghai, and the Kuan-chung Plain of Shensi Province. With a storage capacity of 5.7 billion cubic meters, the reservoir provides flood control and regulation of flow for other projects at Ch'ing-t'ung Gorge and the Great Bend area.

Basic completion of the key project at San-men Gorge was announced on December 20, 1974.[22] After completion of the dam structure in 1960, silt accumulation in the reservoir at San-men Gorge proved so serious that the plans for the project were revamped, and the dam was reconstructed to expand its silt and water discharge capacities. The power plant was redesigned to an installed capacity of 200,000 kilowatts, rather than the originally planned 1 million kilowatts. The most important feature is the reservoir's flow regulation function. Although the huge 64.7 billion cubic meter storage capacity of the reservoir is not utilized under the present plan of operation, vital flood control, irrigation, ice prevention, and municipal and industrial water supply benefits are provided to the lower course of the river.

These announcements quelled speculation outside China that progress on the major dam projects was at a standstill. The silence on dam construction since 1960, together with great emphasis in the Chinese press on other facets of Yellow River management for the last fifteen years, led to the speculation that capital-intensive development had been supplanted by other methods (Chapter 5). These announcements show, to the contrary, that capital-intensive construction remains an integral part of the overall Yellow River development strategy.

It is equally clear, however, that certain very difficult problems were encountered in the construction of the project at San-men Gorge. Since that project is in many ways the most important one of entire scheme, and since it epitomizes the faith put in capital-intensive construction to solve Yellow River problems, it is worthwhile to focus special attention on the history of development at San-men Gorge.

The strong arguments put forward against any construction at San-men Gorge during the first phase of development have been reviewed in Chapter 3. In the 1950s the benefits from such a dam were judged to override those negative features, however.[23] First, the geologic foundation of the dam site was a key consideration since the area is subject to earthquakes. An outcrop of porphyritic diorite underlying the Yellow River channel for a length of only about seven hundred meters at San-men Gorge was considered the most advantageous site by far in that section of the river.[24]

Second, the position of the gorge within the basin made early construction of the dam mandatory in the view of many Soviet and Chinese experts. With more than 90 percent of the basin's drainage area, and most of the major tributaries upstream of this location, the potential for flood control could not be ignored. Many writers agreed that the possibility of significantly reducing or eliminating the age-old problem of Yellow River floods within a few years justified any problems associated with early construction at San-men Gorge. The position of the gorge between Sian, Taiyuan, Loyang, and Chengchow, where industries were expected to develop very rapidly, also argued strongly in favor of early construction.

Third, a reservoir at San-men Gorge would provide true multiple purpose use to a degree that no other reservoir in the vicinity could. In addition to flood control and power generation, the huge San-men impoundment could supply water to many new irrigation schemes, cities, and industries, and could maintain sufficient flow for navigation by five hundred ton steamers.

Against all these advantages was the fact that a reservoir built at San-men Gorge would fill with sediment very rapidly. Reybold, Growden, and Savage had argued that this was sufficient reason never to build a dam at San-men at all, but to build one farther downstream through which silt could be sluiced to the lower course. Cotton had argued that a dam should be built at the key San-men site, but only after the amount of suspended material had been reduced significantly (see Chapter 3).

In deciding to build the San-men Gorge Dam during the first phase of construction, Chinese planners acknowledged that silt accumulation could reduce the volume of the reservoir by 20 percent within the first eight years of its life, and would reduce it by 50 percent within fifty years. Their solution to the silt problem was twofold: a vigorous and complete program of antierosion work on the Loess Plateau, combined with a reservoir large enough to be effective even with large amounts of silt accumulation. Apparently no attention was given to the suggestion that sediment release gates be built in the lower part of the dam, or to the suggestion that silt could be purposely accumulated to the top of one or more dams upstream before construction at San-men Gorge was undertaken.

The design which was adopted for the San-men Gorge Dam specified that the normal high water level of the reservoir would be 360 meters above sea level. This development would allow for both the control of a thousand year flood and the irrigation of forty million *mou*, even after thirty-five billion cubic meters of silt had accumulated in the reservoir. Accumulation of that amount of silt was projected for fifty years after construction was complete.[25]

In choosing to build the 106.5 meter high dam, which put normal high water level at 360 meters above sea level, the designers of the project selected the most expensive plan from among six alternatives.[26] But they considered the costs to be justified by the benefits offered with such a plan. The largest dam required twice the initial capital investment and two and one-half times the number of residents to be relocated, yet it provided nearly twice the generating capacity of the smallest dam and two and one-half times the effective storage capacity. Construction began in April 1957 on the concrete gravity dam that was the showpiece effort of Yellow River management.

The costs at San-men Gorge and other early projects actually went beyond the expenses for materials, labor, construction of new transport facilities, and population relocation. Due to the lack of Chinese industry at the time, heavy earth-moving equipment and other machinery had to be purchased from the Soviet Union, east European countries and Japan.[27] The turbines for the power plant at San-men Gorge were purchased from the Soviet Union, although those for the Liu-chia Gorge power plant reportedly were manufactured in China.[28] Additional expense was incurred at Liu-chia Gorge when the decision was made to put the power station underground. A firm commitment clearly had been made to bringing the Yellow River under control by capital-intensive construction based on heavy state investment.

Beginning in 1957, construction proceeded rapidly at San-men Gorge. Both construction and the relocation of affected villages proceeded ahead of schedule until mid-1960.[29] In July of that year the Soviet advisors withdrew from China, reportedly taking many plans and documents for the San-men project with them.[30] It never has been made clear whether the dam was actually finished to its planned height of 106.5 meters, or whether it remained at somewhat less than 100 meters.[31] During the early 1960s the magnitude of the silt-accumulation problem became apparent, however, and attention was focused on finding a solution to this problem. Power generation equipment apparently was not installed until after reconstruction of the dam was completed.

The reconstruction was carried out in two phases.[32] From 1965 through 1968, two silt discharge tunnels were made around the left end of the dam and eight penstocks were remade into silt discharge tubes. From 1970 through 1973, eight openings in the base of the dam, which originally were used during construction, were reopened. The four remaining penstocks and one penstock which already had been converted to a silt discharge tube were lowered by thirteen meters. Low water head generating equipment, which is resistant to silt abrasion, was installed. The result was an increase in the reservoir's discharge capacity from three thousand to ten thousand cubic meters per second at 315 meters (sea level elevation). These changes not only stabilized the silt accumulation in the reservoir, but also lowered the height of the river channel at T'ung-kuan, thereby alleviating the conditions which were causing siltation of the lower Wei River valley.

This forced compromise in the effectiveness of the San-men Gorge Project reduces it to approximately the level of the least expensive construction alternative of those originally considered. It also is an admission that the silt problem cannot be solved by the methods originally chosen. Erosion prevention on the Loess Plateau simply has not been effective enough to prevent the San-men reservoir's filling at a disastrous rate. Silt is presently being sluiced through the dam, and other measures are being taken to solve the problem. These measures will be discussed in the following chapter.

In spite of the problems at San-men Gorge, however, the recent announcements show that construction at Liu-chia Gorge, Yen-kuo Gorge, and Ch'ing-t'ung Gorge has been quite successful. There are also indications of other successes, although official announcements have not been made. Yellow River officials have stated that five major projects have been completed on the river.[33] It is likely, therefore, that an official announcement may come before too long on another major dam completion. The San-sheng-kung project appears to be the most likely project for such an announcement, since some reports have been made on its progress.[34] There has also been reference recently to three major dams above Lanchou,[35] so it is possible that construction has been proceeding at Lung-yang Gorge, and that this project will be completed within a few years.

It also has been stated unofficially that eleven structures have been completed on the Yellow River, although not all may be in operation at the present time.[36] If diversion structures which do not include dams across the river are referred to by this statement, the projects at Wang-wang-chuang, Lo-k'ou, and Wei-shan might well be included.[37] Another diversion structure which is in at least partial operation is located at Chao-chun-k'an, near Pao-t'ou.[38] It is not known whether this diversion is connected to the early, unsuccessful effort on the Salachi Irrigation District (Chapter 3).

Beyond this the paucity of information on specific Yellow River projects makes it impossible to speculate which structures may be completed or nearly completed. *Ajia no Yume* reported that the dam at Hei-shan Gorge, below Lanchou, not only was begun in 1958, but was completed in 1960. This cannot be confirmed, however, without information from Chinese sources. In spite of the fact that the T'ao-hua-yu project was originally planned for the initial phase of development, no construction apparently has taken place on this project. A dam was originally planned above the railroad bridge near Chengchou. When it was realized that reduction of the river's silt load by the San-men Gorge project would cause shifts in the river channel, however, construction was delayed and plans were made for the dam to be built downstream from the bridge.[39] Construction on a dam at Hua-yuan-k'ou below the bridge reportedly began in 1959,[40] but visits to Hua-yuan-k'ou and the Mang-shan Pumping Station above the bridge in late 1974 showed no evidence of such a dam or reservoir.[41]

In spite of the questions remaining about the status of many projects, the announced completions of four major dams on the Yellow River make it quite clear that capital-intensive construction by the central government remains a very important component of current Yellow River development strategy. It is not the exclusive method of development, but neither has it been de-emphasized or curtailed as the information from China after 1960 tended to suggest. The other methods, which have been combined with capital-intensive construction, will be described in Chapter 5, but first it is important to appraise the agencies which evolved to implement the Yellow River scheme during the decade of the 1960s.

Operation of the Yellow River Water Conservancy Commission

The Yellow River Water Conservancy Commission (YRWCC) (Huang-ho Shui-li Wei-yuan-hui) was created in 1933 to coordinate water management efforts throughout the Yellow River basin. Li Yi-chih and others had argued that the creation of a strong basin-wide agency which transcended provincial governments in authority, and which had direct links to the central government, was the only way to effectively implement basin-wide control of the river. During the 1930s and 1940s the Commission coordinated most of the surveying and preliminary planning work that was completed, but it did not gain the degree of autonomous authority which its early proponents had hoped for.

During the 1950s, with the vast expansion of development work in the Yellow River basin, the YRWCC assumed many more responsibilities. Figure 4-1 depicts the organizational structure of the Commission as compiled from various references to organizational relationships appearing in the Chinese press. The Commission convened numerous conferences on topics ranging from flood preven-

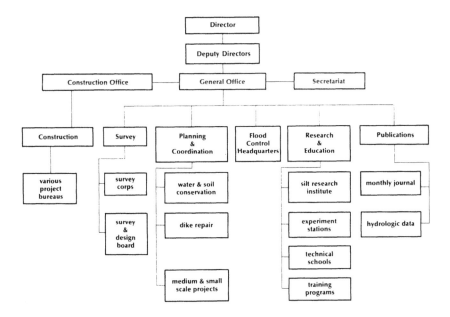

Figure 4–1. *Organizational Structure of the Yellow River Water Conservancy Commission*

tion to water and soil conservation, and from navigation to power generation. These conferences were attended by national, provincial, county, and commune leaders, as well as by technical, scientific, and trade union personnel. Research, education, and publication activities also constituted major parts of its work.

Implementation of management plans is not a primary task of the Commission, however. Certain reports indicating conflicts between the YRWCC and the Ministries of Forestry and Agriculture illustrate the point. In water and soil conservation work the YRWCC was said to have responsibility, but little delegated power.[42] The Commission emphasized soil conservation work, while the Ministry of Forestry seemed interested only in afforestation, and the Ministry of Agriculture concentrated on improving farming techniques. Not only was coordination between the programs lacking, but the internal organization of each of the agencies involved was also criticized as being ineffective. It was further charged that many cadres, as well as peasants, did not understand the relationship between water and soil conservation and agricultural production, but rather viewed the two activities as competing for resources and the peasants' labor.

Other evidence of a serious lack of interagency planning and coordination is apparent in the treatment of reclaimed lands. Reclamation projects were opened for a season or two and then the newly cultivated fields were abandoned and left completely vulnerable to erosion. Such practices brought the sharpest and most persistent complaints from those involved with soil and water conservation.[43] In other instances the control of forests was in dispute, and, with no management or protection, timberlands were excessively exploited.[44]

When soil and water conservation plans had been agreed upon, the method of implementation still was contested. Some advocated vegetative solutions (planting trees, shrubs, and grass cover), some advocated engineering solutions (building check dams and terraced fields), and some advocated farming solutions (contour plowing, etc.).[45] A more unified method of administration and implementation clearly had to be developed.

Administrative problems arose because the YRWCC superseded provincial governments, but it came into conflict with other agencies of the central government. The YRWCC did not have the power to oversee and give primary direction to all facets of development in the basin. In the construction and operation of major dam projects, the YRWCC had even less authority. Its various project bureaus were little more than advisory adjuncts to project bureaus of the Ministry of Water Conservancy or the State Council. Thus the YRWCC functioned more as an interagency coordinating unit than as a strong and autonomous executive unit for basin development.

The actual center of administrative control for the Yellow River development scheme was located in the central government, in a staff office of the State Council. The State Council is the primary executive body of the central government in China. The various ministries through which the government operates are overseen by staff offices of the State Council.[46] The relationship of the various agencies responsible for development of the Yellow River basin is shown in Figure 4-2. The YRWCC is shown to coordinate efforts among numerous agencies at many levels. Another unit, the Yellow River Planning Commission, did much of the basin-wide planning work in the 1950s, although many of its functions may have been assumed by the YRWCC in the last fifteen years.

The insufficient amount of information on Yellow River development since 1960 makes it difficult to evaluate the directions in which the water management agencies have evolved. From the reports available, however, it seems clear that authority remains with the central government for overall planning and major project development. The YRWCC continues to function as a coordinating agency.

Increasing emphasis on the principles of local initiative and self-reliance, discussed in Chapter 5, has great significance for the agencies involved in Yellow River development. Local agencies assume more responsibilities and gain more importance. Within broad guidelines specified by the central government, provincial, county, municipal, and commune agencies plan, construct, and operate their own projects. Provincial governments are responsible for large projects, such as the flood prevention levees on the lower course of the river and major irrigation schemes; county or municipal governments are involved with large water supply or irrigation schemes; commune water conservancy agencies are responsible for diversion or storage projects within the boundaries of the commune.[47]

The flexibility in water management decision making which this system affords is a very positive feature. If all decisions were made at or above the basin level, such flexibility would not be possible. Technology has played a part in enhancing this flexibility, of

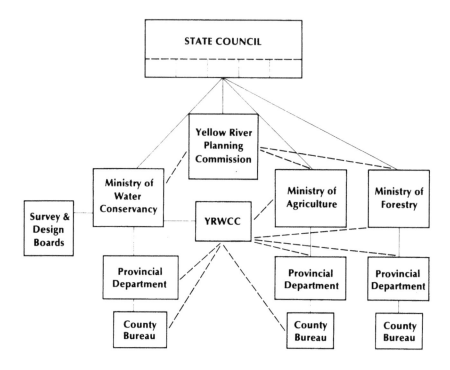

Figure 4–2. *The Yellow River Water Conservancy Commission and Related Governmental Organizations*

course; the development of pump and siphon technology since the Yellow River scheme was first announced in 1955 has facilitated local initiative and self-reliance. But the reliance on local agencies for a great degree of basic decision making is an even more striking characteristic of recent Yellow River development strategy. It is a feature of water resources management policy which is recognized as being increasingly important.[48]

In summary, the decade of the 1950s was an extremely important period in the history of Yellow River management. The early use of scientific techniques was consolidated and expanded during this period, and a basin-wide development plan was implemented for the first time. Several key dam construction projects were under-taken with Soviet assistance, some of which have recently been completed. Agencies capable of implementing the basin-wide plan were established, and these have continued to evolve in recent years. In addition to utilizing new methods for Yellow River management, however, other methods, which are reminiscent of historical Yellow River control strategies, were also used. The adaptation of historical methods to modern conditions is perhaps the most significant feature of current basin management.

Chapter 5
Reliance on Traditional Methods

While important new water management techniques were introduced into the Yellow River basin during the decade of the 1950s,
traditional techniques were by no means abandoned. Scientific
methods, large dam construction, and other modern techniques
were introduced, but reliance on labor-intensive construction also
continued. In the maintenance of flood levees and the construction of irrigation canal systems, the mobilization of large labor
forces continued as it had through centuries of Yellow River control. Together these efforts typified the Chinese principle of "walking on two legs," combining native and foreign methods.

Mass Labor Mobilization in Modern Water Management

The flood prevention levees on the lower course of the Yellow
River demanded considerable attention during the 1950s, because
they had not been adequately maintained during the years of war,
and because high water crests continued to rise every flood season.
The work was undertaken with enthusiasm, however, because a new
era in flood prevention was anticipated. The large dams and antierosion works planned for completion in the middle basin during
the 1960s would ultimately solve the age-old flood problem. Maintenance of the levees downstream during the 1950s was described
as temporary, therefore, necessary only until upstream projects
were completed.[1]

The traditional method of levee maintenance work was utilized.
Large numbers of agricultural laborers were mobilized in the slack
season to carry out earth-moving operations. A winter repair (tung-
hsiu) operation typically is organized in November to early December, and a spring repair (ch'un-hsiu) operation in February to

March. The scale and scope of dike-strengthening projects is indicated in Table 5-1. While this information, obtained from scattered Chinese press sources, does not provide a precise and detailed description of yearly progress, it does give a general picture of levee maintenance work.

Other kinds of mass labor mobilization for flood protection include channel dredging and levee protection through the planting of trees and grass. Very little information is available on dredging, but one report in 1958 describes the mobilization of ten thousand workers to remove deposited material from the channel in Shantung.[2] One million saplings were planted on the dikes to protect them from erosion, in the spring of 1954, 2.2 million were planted in the winter of 1961, and 3.6 million in the winter of 1963. Five million square meters of grass were planted on the dikes in the spring of 1954, and 4.4 million square meters were planted in the winter of 1963.[3]

The mobilization of labor was greatly facilitated by the collectivi-

Table 5-1. *Representative Dike-Strengthening Projects in Honan and Shantung Provinces*

Project	Labor Force	Material in Million m³ earth (m³ stone)
Spring 1950	100,000	
Spring 1952	100,000	6.5
Spring 1953	66,000	
Spring 1954	50,000	10
Winter 1954	70,000	2
Spring 1955	180,000	32 (360,000)
Spring 1956	160,000	17 (60,000)
Winter 1956		4.6
Spring 1957	120,000	8.8
Winter 1957	75,000	3.9 (34,000)
Winter 1961		.23 (Shantung only)
Winter 1962	150,000	
Winter 1963		7.1 (40,000)
Winter 1964	70,000	

Source: Various NCNA reports from 1950 through 1964.

zation of agriculture. The mutual aid teams, agricultural cooperatives, and later the communes were the units responsible for implementing labor goals (e.g. three cubic meters of earth moved per
day per worker), and for motivational techniques. Since 1958 the
production teams and production brigades that comprise the communes have provided the basic units of labor mobilization.

Certain aspects of flood prevention work appear not to have
changed to a great extent under the new government. The primary
task remains unchanged; it is to move large amounts of earth and
stone for levee construction. The primary tools for earth-moving
work are still wheelbarrows and baskets on shoulder poles. Pay-
ment for laborers still is given in kind, either as grain or as work
points, rather than in money wages. Provincial River Affairs Bureaus
(Ho-wu Chu) in Shantung and Honan continue to take primary
responsibility for the winter and spring projects as they have for
centuries, albeit under the general coordination of the YRWCC.

Similarly, the basic problem of flood control has not been removed. Two flood crests within the first decade of the Communist
period were among the highest of this century. On September 4
and 5, 1954, a flow of 15,000 cubic meters per second was recorded
at Shan-hsien, and on July 17, 1958, the flow reached 21,000 cubic
meters per second at Hua-yuan-k'ou near Chengchou. Both were
well above the normal flood crests of 5,000 to 8,000 cubic meters
per second. The 1958 crest, in which the levees were breached in
five places, approximated the disastrous flood of 1933 (23,000 cubic
meters per second at Hua-yuan-k'ou) during which the dikes broke
in more than fifty places.[4]

Although the amount of damage resulting from the dike breaches
is not reported in the Chinese press, it is clear that from 1949 on-
ward the flood control and prevention measures began steadily to
reduce the amount of damage caused by floods; but the problems
remained complex. Not only were discharge amounts high at times,
but also the age-old problem of silt was still apparent. It was reported in 1956, for example, that a peak flow of 8,000 cubic meters
per second reached a higher level on the dikes than the 1954 crest
had reached, even though the flow volume was about half, because
silt deposited during the winter had not yet been scoured by spring
discharge.[5]

Ice floes in the winter pose another serious problem to maintenance of levees. When the ice breaks up it floats downstream and
occasionally jams up, blocking the river channel and threatening

the levees. Explosives and bombs have been used in the past to break up these ice jams, but reports indicate that as late as 1969 they were still a significant problem.[6]

The problem of floodplain channel stabilization also remains, but in it lies a perspective on the changed management conditions in the Yellow River basin. No longer does the fundamental question of Yellow River control revolve around how far apart the levees should be built, or what the effects of flow and scour will be. These simply are problems to be studied and solved as part of a larger scheme. Whereas the floodplain channel was the major focus of traditional management strategy, in the modern, basin-wide strategy, it is considered in relation to reservoirs, soil conservation projects, and other measures.

Between 1949 and 1953, 800 kilometers of flood prevention levees were strengthened.[7] By August 1955, the total had reached 1,800 kilometers; 130 million cubic meters of earth and stone material had been moved.[8] By the same date, fourteen million saplings and sixty-six million shrubs had been planted. A report in 1965 indicated that the primary emphasis in dike-strengthening work continued to be maintenance of the Yellow River levees, although dikes around flood detention basins also received some attention.[9] The main levee on the north bank of the river is seven hundred kilometers long, starting near Meng-hsien in the west. On the south bank starting near Chengchou, the main levee is more than six hundred kilometers long.

Smaller flood prevention dikes also are necessary in various tributary basins and on the mainstream in the middle reaches. In the Meng-ho basin in western Honan, for example, flood prevention dikes were constructed as part of a multiple purpose plan to develop the basin's resources.[10] Similar levees are maintained in the Wei, Fen, North Lo, South Lo, and other basins.

Various flood prevention levees are maintained along the Yellow River in Inner Mongolia. The construction of the levee at Salachi Hsien, which is ten *li* (three miles) long, required six thousand laborers to move 600,00 cubic meters of earth in 1956.[11] Similarly, on the Ninghsia Plain the annual repair of river levees required thirty thousand workers in 1962; in Kansu, smaller levees in the vicinity of Lanchou also required regular maintenance.[12]

Along this lower course of the Yellow River, flood diversion works and detention reservoirs were constructed to supplement levees during the early years of the Communist government. By

1955, three major diversion projects operating together could successfully accommodate a flow of 29,000 cubic meters per second (a flood statistically predicted to occur only once every two hundred years). These major diversion works are the Tung-p'ing Lake Reservoir, the Shih-t'ou-chuang overflow wier, and the Ta-kung diversion works.

Tung-p'ing Lake in western Shantung is the largest detention basin on the Yellow River. Water is diverted at the Wei-shan multiple purpose dam near Tung-a County, some eighty kilometers southwest of Tsinan. Initial dike construction was completed in 1951 on the Tung-p'ing Reservoir, and improvements were made continuously thereafter. During the flood crest of mid-July 1958, a total of seven hundred million cubic meters of water were diverted into the lake, reducing the flow by more than two thousand cubic meters per second in the main Yellow River channel. Plans were announced later the same year to double the detention capacity of the reservoir by raising its dikes.[13]

The Shih-t'ou-chuang overflow weir is located at Ch'ang-yuan Hsien, seventy-five kilometers downstream from Kaifeng. It is located in the "soft midsection" of the Yellow River where the levees breached most frequently in the past. An opening some 1,500 meters long was made in the main north levee, faced with stone, and blocked by a "control dike" that can be opened to divert flood waters. Diverted water flows between the north bank levee and another abandoned levee, eventually returning to the Yellow River more than thirty kilometers downstream. The project was completed in three months during the summer of 1951 by a corps of forty-five thousand workers, and involved moving 250,000 tons of material. The increase in soil fertility from silt deposited by the slowed flood waters is a main benefit of this project.[14]

The Ta-kung flood diversion project also diverts flood water into an area between two dikes on the north side of the Yellow River. It is located near Feng-ch'iu across the river from Kaifeng. Completed by sixteen thousand workers from May to June 1956, it was designed to divert six thousand to ten thousand cubic meters per second of water when the flow is thirty-six thousand cubic meters per second.[15] (Such a flood is statistically predicted to occur only once in one thousand years on the Yellow River.)

Another striking example of labor mobilization is the "flood fighting" work of the Yellow River Flood Prevention Headquarters, organized by the YRWCC. The headquarters holds an antiflood conference in late spring every year to lay plans for the approaching flood season, during which as many as ten flood crests may threaten the levees. A network of more than three hundred stations in all parts of the Huang Ho basin reports precipitation and discharge amounts throughout the flood season, usually from June 1 through October 31, to facilitate the prediction of the place and time that flood crests may be expected.[16]

When the river rises to threatening levels, rescue teams and flood emergency workers, organized by provincial governments but trained and coordinated by the Flood Prevention Headquarters, are summoned. Rescue teams include about sixty workers, cadres, and engineers who are equipped with vehicles, generator-operated search lights, and other emergency equipment. Emergency work teams are stationed on vulnerable sections of the dikes and perform immediate repairs on breaches of the levees. It was reported that as many as three hundred workers per five hundred meters were stationed on certain sections of the levees in Shantung during the high crest of July 1958.[17]

The completion of the San-men Gorge Dam has greatly reduced the threat of flood waters that originate on the Loess Plateau and farther upstream. Even with effective regulation at San-men Gorge, however, the threat of flood waters arising in the Ch'in, Southern Lo, and other tributaries below San-men (which were the origin of much of the 1958 crest) remains ever present.

Reduction of the flood hazard has not eliminated the need for flood control programs and the need for vigilance against levee breaches. The Yellow River Flood Prevention Headquarters continually must guard against overoptimism about the effects which new diversion works, conservation projects, or dams will have in reducing the flood hazard.

All these features make flood prevention the single most successful aspect of the entire Yellow River scheme to date. No disastrous floods, of the kind which historically devastated the North China Plain, have occurred since 1949. Flood prevention is the most basic necessity for the overall management plan; without it the stability and prosperity necessary for other development do not exist. The method used to attain this success is also the most basic one; mass mobilization of labor forces for levee construction is the main legacy of traditional Yellow River management methods in modern times. Labor mobilization has been used in irrigation and soil conservation work since 1949, but the results in these two spheres have not been as substantial as in flood prevention work.

The problems which irrigation workers confronted in 1949 were numerous and difficult. Historically, diversion of the Yellow River for irrigation on the floodplain threatened the security of the levee system. The frequency with which the river shifted its channel and the difficulty of keeping the levees strong made large-scale irrigation diversions treacherous. The utilization of modern technology to control the maximum discharge, to stabilize the channel, and to construct modern sluice gates increased the possibility of greatly improving irrigation on the floodplain, however.

A variety of poor soil conditions also confronted the early irrigation plans. Much of the land on the floodplain had become depleted of nutrients. Large areas, particularly near the coast, were poorly drained and so highly alkaline as to make the soils unproductive. In other areas of the North China Plain sandy wastelands existed in abandoned Yellow River channels.

In the middle basin existing irrigation works in Ninghsia, the Great Bend region, and the Wei basin also faced numerous problems. In spite of the progress prior to 1949, many canals were badly silted, much land was waterlogged from poor drainage, levels of alkalinity were excessive in some areas, and the existing diversion gates were in very poor condition.

The solution to these problems in the early 1950s was the construction of irrigation and reclamation projects. During the decade of Soviet assistance several large projects were designed and partially implemented. The basic means of development was labor mobilization for earth-moving operation, but capital investment was also contributed by the state for building sluices, headgates, and other costly features.

The People's Victory Canal was the first irrigation project opened on the floodplain. Inaugurated in early 1952, it marked "the first time in history that Yellow River water was used to irrigate wheat farms."[18] Forty cubic meters per second of water were diverted from one mile west of the Peking-Wuhan railroad bridge near Chengchou. The canal flowed into the Wei River near Hsin-hsiang, fifty kilometers to the northwest (see Map 5-1). The irrigated area was 300,000 *mou* in 1952, and this amount was doubled by 1955. The People's Victory Canal was only the first stage of the much larger Yin-huang Chi-wei ("Divert the Yellow River to Benefit the

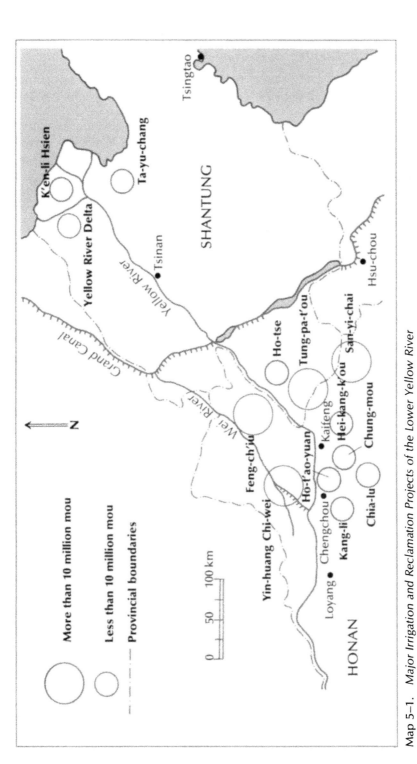

Map 5–1. *Major Irrigation and Reclamation Projects of the Lower Yellow River*

Wei River'') project, and it signaled the large-scale construction methods subsequently developed elsewhere on the plain (see Tables 5-2 and 5-3).

The Yin-huang Chi-wei and Ta-yu-chang projects were designed to bring water to the largest alkaline areas near the coast. A reclamation experiment station was established in 1953 south of the Yellow River in the Ta-yu-chang project area. Based on the findings of this and other similar stations, large areas of alkaline soils were reclaimed. Water from the Yin-huang Chi-wei Canal flowed into the Wei River, then through the Grand Canal to the vicinity of Tientsin, where coastal marshes in Hopei Province were washed of

Table 5-2. *Planned Area of Major Lower Yellow River Irrigation and Reclamation Projects (Area Given in Million Mou)*

A. Canal Projects

Project	Planned Irrigated Area	Reclamation Area	Diversion Rate (cms)	Paddy Rice Area	Canal Length (km)
Yin-huang Chi-wei	15	2.4	280–350	2.4	115
San-yi-chai	21	1.6		13	
Ta-yu-chang	5.12	1.4			22,000
Ho-t'ao-yuan	0.2	0.047			
Hei-kang-k'ou	1	0.1	50		
Kang-li	7	0.8	280	4	
Feng-ch'iu	11				
Chia-lu	2.5				
Chung-mou					160

B. Irrigation Reservoirs

Project	Irrigated Areas	Storage Capacity (million cu m)
Tung-pa-t'ou	19	1,700
Ho-tse		950
Yellow River Delta	3.4	2,200

Source: Various NCNA reports from 1952 through 1960.

Table 5-3. *Construction Progress on Major Lower Yellow River Irrigation and Reclamation Projects (Area Given in Million Mou)*

Project	Irrigated Area	Date Completed	Million Cu M of Earth	Labor Force
Yin-huang Chi-wei	1.4	1961	26	200,000
San-yi-chai	1.3	1958		220,000
Ta-yu-chang	5.12	1958	18	220,000
Ho-t'ao-yuan	0.2	1956		
Hei-kang-k'ou	1.0	1957		
Kang-li	7.0	1958		
Feng-ch'iu	11	1958		225,000
Chia-lu	2.5	1961		
Chung-mou Tung-pa-t'ou	19	1958	23	150,000
Ho-tse			41	100,000
Yellow River Delta	0.5	1959		
K'en-li Hsien	0.7	1966		

Source: Various NCNA reports from 1957 through 1966.

harmful concentrations of alkaline salts. In the Ta-yu-chang project Yellow River water was diverted east of Tsinan to flow through the Hsiao-ch'ing River to the coastal counties of Po-hsing and Kuang-jao.

A third coastal reclamation area, on which there is very little information available, is located in the midst of the Yellow River delta itself. K'en-li ("Reclamation Benefit") Hsien, consisting of six communes, three state farms and a forestry station, was established here in the middle 1960s. The population of the new county was 150,000, and the reclaimed area was 700,000 *mou* in 1966.[19]

The Ho-t'ao-yuan project is representative of the efforts to reclaim abandoned river channels. With a diversion point near the 1938 dike breach at Hua-yuan-k'ou, it was designed to reclaim the land wasted by that breach.

A comparison of Tables 5-2 and 5-3 shows that the areas planned for irrigation in the largest projects had been developed only fractionally by the late 1950s. Although some smaller projects were said to be complete, the figures given in that period were later shown to be unreliable. Of unquestioned significance for later development were the diversion structures actually completed, however. Fourteen major sluice gates had been constructed, twenty-two pipes had been driven through the levees, and forty-four siphons were in operation by 1959.[20] The largest sluice gate (410 meters long) was on the San-yi-chai project on the Honan-Shantung border.

The development of paddy rice cultivation in the lower Huang Plain was an important innovation of the new irrigation schemes. Based on the reliable supply of irrigation water, paddy rice was introduced into this area where it had not been grown before. The degree to which this program has been successful is difficult to ascertain, however, since only scattered references to rice production in the Yellow River basin have been made.

In addition to their primary benefit of providing irrigation water, these schemes included important secondary benefits. Use of the canal system for inland shipping is one example. The People's Victory Canal, and later the larger works of the Yin-huang Chi-wei project, for example, guarantee inland navigation for two hundred ton ships for a distance of nine hundred kilometers from Chengchou via the Wei River and the Grand Canal to Tientsin. South of the Yellow River, the Hei-kang-k'ou and Chia-lu river projects provide shipping links from Kaifeng and Chengchou, respectively, to the Huai basin.

Hydroelectric power generation is another secondary benefit. Small generating plants were built on several of the canal systems, with the power generally going to electrification of pumping stations and other rural uses. Three of four such stations planned on the Yin-huang Chi-wei project were in operation by 1961, for example, and four stations were completed on the Ta-yu-chang system by early 1960, with a total generating capacity of 2,600 kilowatts. No report has been made on the planned generating capacity of 1,280 kilowatts on the San-yi-chai system or 1,200 kilowatts on the Tung-pa-t'ou system.

Irrigation districts in the middle portion of the basin also were improved through mass labor mobilization during the 1950s. In the Ho-t'ao district the "Liberation Diversion Gate" (formerly called the Huang-yang Sluice Gate) was constructed in 1952 to divert water to three trunk canals at the western end of the ancient irrigation system. The eastern part of the district and the south bank of the river subsequently received water through other construction projects. Between 1959 and 1961, in connection with the San-sheng-kung Dam, the "Second Yellow River Canal," 190 kilometers in length, was built, north of and parallel to the Yellow River. Thirty million cubic meters of earth were moved so that the ten main canals of the irrigation system could be fed from this waterway, rather than directly from the Yellow River. This canal, plus 400 kilometers of flood prevention dike, has greatly reduced the flood threat not only in the canal system, but also in the Wu-chia River and the Wu-liang-su Lake which drain the system back into the Yellow River.[21]

Another major undertaking in the Hou-t'ao Irrigation District was the adjustment of canals to accommodate the Pao-t'ou–Lanchou Railroad bed. One and one-half million cubic meters of earth were moved from November 1956 through June 1957 to change the canal pattern so that the number of bridges and culverts required by the roadbed was reduced from eight hundred to fifty-eight.[22]

In the Ninghsia District renovation of the canal system proceeded during the 1950s, prior to completion of the Ch'ing-t'ung Gorge Dam. The dam, which is located at the head of the Ninghsia Plain, is near the diversion point of the ancient irrigation canals. Between 1953 and 1955, eighty-four kilometers of the Han-yen Canal, for example, were renovated to irrigate 780,000 *mou* near Yin-ch'uan; forty thousand laborers were reportedly mobilized in 1956 for a concentrated effort on spring canal repair; and in the winter of 1959, the Hsi-kan Canal was repaired to raise water through a series of dams to irrigate 150,000 *mou* at high elevations near the Holan Shan.[23] The Ch'ing-t'ung Gorge project began diverting water for irrigation in 1960, and by 1963 the area reportedly under irrigation was three million *mou*, double the acreage irrigated in 1949.[24]

Permanent flood prevention levees were also constructed in a mass labor campaign in Ninghsia during 1963–1964. A total of 288.2 kilometers of levees were constructed on flood-prone stretches along both banks of the river Chung-wei to Shih-tsui-shan.[25]

Soil and water conservation work on the Loess Plateau was another part of the overall scheme in which heavy reliance was placed on mass labor mobilization campaigns. In spite of the general agreement that conservation work (or, as it often appears in Chinese publications, "antierosion work") would have the most lasting impact, it was not given the highest priority in the early phase of construction. The reason was spelled out as early as 1955, when

Teng Tzu-hui first made public the basin-wide plans: "Obviously, it will take decades to carry out the entire scheme. For instance, the part of the scheme for conservation of water and soil alone will take fifty years. So, because we must solve first things first—the problem of how to prevent the Yellow River flooding, and how to use the river for power generation and irrigation, and other urgent problems—the Yellow River Planning Commission is submitting a 'first phase plan'—part of the comprehensive plan—to be carried out within the period of three five year plans, that is, before 1967."[26]

Lip service was given to retarding the loss of soil and water resources through erosion, but the goals were only superficially fulfilled during the 1950s. The failure of check dams and other erosion prevention works within a year after they were constructed was common during this period. Often after a check dam or other feature was completed it was not maintained properly, so that it filled with silt or was washed out in a few months' time. The clear implication is that in some quarters there was more concern for statistical success than for actual accomplishments.[27]

The lack of firm leadership at all levels of operation was another frequently voiced problem; it is clear that the tasks of educating the peasants to the importance of antierosion work and of mobilizing them were not accomplished. The interagency conflict regarding conservation work, discussed in Chapter 4, was also a major impediment to progress.

The first step toward solving these problems was taken at a conference called in the fall of 1959 by the Committee on Soil and Water Conservation of the State Council.[28] A higher priority was assigned to antierosion work at the conference, but there is no report of how new plans were to be implemented. Although reports of the interagency conflicts do not reappear in published materials, neither is there any mention of a single ministry or agency having been given primary responsibility for soil conservation work.

The increased attention given to antierosion work seems to have been forced by circumstances, rather than having been initiated through foresight. It is stated outright that the conference was prompted by the near completion of the San-men Dam and the realization that the reservoirs behind the dam were already accumulating vast amounts of silt.[29]

The method for accomplishing basic control of erosion on the Loess Plateau was reliance on the masses. Ten million workers were reported to be mobilized in the autumn of 1959 in the seven provinces and autonomous regions of the middle basin. It was by far the most concerted effort at antierosion work undertaken up to that time.

No comprehensive results of this mass project have been published, but continued discussion of the soil and water conservation problems through the 1960s indicates that the struggle was a continuing one. From 1963 through 1965 the efforts and accomplishments were reported each year as the biggest ever in one province or another.[30] Although efforts were redoubled in every province and autonomous region, the most intense activity was concentrated in northern Shensi and western Shansi, the area of most severe erosion.[31]

More than any other aspect of the modern Yellow River project, therefore, soil and water conservation work increased its reliance on mass mobilization during the 1960s, with very little capital investment coming from the central government. State aid has played a minor role, as indicated by the following statement:

> In order to speed up the construction of the mountainous areas and the harnessing of the Yellow River, appropriate aid from the State is also necessary. But this kind of aid must be built on the foundation of relying on the collective strength of the masses. All projects that can be undertaken by communes and teams must be undertaken by them. When a project is beyond the power of one production team, it should be jointly undertaken by several production teams and undertaken by them in cooperation with the production brigade. State money can be used only for propaganda, rewards, scientific experiments and the purchase of seeds and saplings and to help finance the comparatively large engineering projects which may be too difficult for the teams to undertake. It is only in this way that water and soil conservation can be turned into the conscious action of the broad masses.[32]

Reports since the Cultural Revolution indicate that the strategy of relying on the masses has continued until very recently. To a large degree it is still the "broad masses of poor and lower middle peasants" who bear the burden of soil and water conservation work.[33]

Recent labor-intensive methods represent considerable modification of earlier methods, however. Irrigation and soil conservation work proved less successful during the first decade of Communist government than did levee construction work. Problems arose in irrigation and conservation work which required changes in the water management strategy. The labor-intensive methods which were quite successful in flood control works encountered problems in irrigation and antierosion work.

Such problems should be placed in the context of the general climate for development in China after a decade of Communist rule. Following the enthusiasm of the Great Leap Forward, a more sober outlook began to emerge. Withdrawal of Soviet assistance in 1960 and the "bitter years" of drought and natural disasters, from 1959 to 1961, took their toll on the physical and psychological basis for development.

The problems in the Yellow River scheme cannot be attributed solely to this climate, however. Specific management programs in soil conservation and irrigation proved to be very unsatisfactory. The clearest evidence of problems in soil conservation work is the recent announcement of silt accumulation in the San-men Gorge Reservoir. It has been acknowledged that a silt layer nearly twenty meters thick had accumulated in front of the dam at San-men Gorge, and that serious problems arose from deposition near the mouth of the Wei River (see Chapter 4). This problem has been blamed on miscalculations made while the Soviet advisors were involved in the Yellow River project.[34] Wherever the blame may lie, it had become clear by the early 1960s that the silt problem could not be solved as easily as originally hoped.

Other evidence shows that problems also arose with the huge irrigation canal systems on which construction with mass labor mobilization had begun during the 1950s. Irrigated acreage did not continue to expand during the early 1960s and the evidence indicates that soil alkalinity and satisfactory methods of rice cultivation had not been mastered. The few reports available describe campaigns of continuing struggle with these problems.[35]

The Ta-yu-chang and People's Victory Canal projects are mentioned as specific examples of irrigation program failures. Wheat production had been increased by 100 percent in the early years of the Ta-yu-chang project, but it decreased during the early 1960s. Such failures provided ammunition for later attacks on the strategy of development through large project construction. After the Cultural Revolution the blame was placed on "Liu Shao-chi and his counter-revolutionary agents," for relying "on a few specialists to do designing in isolation from the masses. They paid attention to the construction of only big sluices and trunk canals to lead water from the Yellow River, but neglected to mobilize the masses and rely on them to build branch canals, ditches and other auxiliary projects. And they paid attention only to irrigation which was done by flooding to the neglect of drainage." [36]

Similar attacks were made against those who used the same strategy on the People's Victory Canal. They had advocated the method of "large-scale drawing of water, large-scale storage, and large-scale irrigation in the irrigation area, and made indiscriminate irrigation without making drainage. This caused the drastic elevation of the underground water table in the irrigation area and resulted in the alkalinization of farm fields, aggravating the calamities of waterlogging and alkalinization at one time." [37]

By the time these attacks were published (1971), new measures had already been implemented for several years to solve the problems described. The labor-intensive methods for irrigation and soil conservation work had been considerably modified to solve the specific problems of different areas. The modifications were part of a larger shift in Yellow River management strategy occasioned by changes in national development policy.

Local Initiative and Self-Reliance
in Solving Local Problems

Beginning in 1957, a series of planning decisions was made which changed the emphasis of water management work throughout China. The decision regarding farmland irrigation projects was issued in September 1957 by the Party Central Committee and the State Council. It stated that primary emphasis should be given to small-scale irrigation projects, that secondary emphasis should be given to medium-scale projects, and that large irrigation projects should be built only when "necessary and feasible." [38] The directive further stated that the masses were to be relied upon for labor and that local capital was to be used to finance the projects.

In August 1958, this policy was extended beyond irrigation work by the Party Central Committee's "Directive on Water Conservancy Work." With this directive the "Three Primaries" campaign was launched, and primary emphasis in all phases of water management was placed on small-scale projects, on water storage, and on the newly formed people's communes as units responsible for project construction.[39]

Subsequent national development policy decisions continued the same trend,[40] and the effect of all these guidelines was to shift the emphasis from large, capital-intensive projects, toward small projects undertaken with local resources. During the 1960s the main water management principle enunciated was "multiple purpose control in light of local conditions." In Yellow River management, therefore, primary attention was focused on small- and medium-scale projects that combined flood prevention, irrigation, and drainage benefits in a manner that best solved local problems.[41] The recent announcements on major dam completions demonstrate that large-scale capital-intensive construction was not entirely abandoned during this period, but the major emphasis clearly was shifted to constructing numerous small projects rather than limiting development to the major "staircase" dams on the Yellow River mainstream.

Application of these principles to the major irrigation schemes on the plain resulted in an increased number of small diversions from the Yellow River. The large canal systems built during the 1950s have been maintained and continuously extended; but rather than being supplied from a few diversion points with large sluice gates as originally planned, the canals are fed from numerous smaller pumping stations, siphons, and small sluice gates. In late 1974 a total of sixty-five sluice gates and eighty pumping and siphon stations was reported on the lower Yellow River.[42]

Siphon stations are the simplest and most economical of these diversions to install. The siphons draw irrigation water over the levees from the elevated channel of the Yellow River with relative ease and very inexpensively. This program is most widely developed in Shantung, where every county along the river has constructed siphon systems. In Li-ch'eng Hsien northeast of Tsinan and other counties in the delta region, siphon irrigation has been used to reduce soil alkalinity and to irrigate alkaline-resistant strains of paddy rice.[43]

Pumping stations represent a unique combination of capital-intensive and labor-intensive methods for irrigation development by local units. The Mang-shan Pumping Station in western Honan is a good example. Undertaken by the city of Chengchou, the Mang-shan Station serves an irrigation area of 100,000 *mou* and provides a portion of the urban and industrial water used by Chengchou.[44] The main capital component of the project is the pumping equipment, consisting of sixteen 20-inch-diameter pumps and two smaller pumps. The labor component consisted of four million labor days for construction of the canal system, tunnels, and aqueducts. Maintenance and further development of the project likewise utilize a combination of labor and capital inputs.

The use of well irrigation also increased rapidly during the 1960s. The electrification of pump wells spread rapidly in this period, and by 1974, well irrigation accounted for 15 percent of the total irrigated area in the Yellow River basin.[45] An extensive system of wells has been established to improve the effectiveness of the People's Victory Canal, for example. Numerous wells were opened in this area between 1965 and 1970 to lower the water table, reduce alkalinity, and make irrigation more effective.[46] Pump well projects can be undertaken by a commune- or production-brigade-sized unit. They represent the lowest extension of the strategy of combining capital-intensive and labor-intensive methods to fit local conditions.

In the middle portion of the Yellow River basin the 1960s were similarly a time of stressing medium- and small-scale projects undertaken with local initiative. Small projects had received greater emphasis during the 1950s in this area than they had on the plain downstream. It was reported in 1959, for example, that of thirty thousand reservoirs on all of the Yellow River tributaries only twenty-nine had dams forty meters or higher and were capable of irrigating more than 500,000 *mou*.[47] In the same year a total of 4.3 million *mou* were irrigated by two thousand small projects on streams and mountain torrents in a few counties of southwestern Shansi.[48] Undertaken by a single commune or production brigade without state assistance, such projects became increasingly numerous after 1960.

During the decade of the 1960s the combination of labor-intensive and capital-intensive methods was used both to open new projects and to improve irrigation work which had been completed unsatisfactorily during the 1950s. In Ninghsia, as along the lower portion of the river, construction on the major canals during the 1950s had not been matched by construction on small canals for both irrigation and drainage.

In the "Seventy-two Connected Lakes" area of Ninghsia, for example, a total of 3,900 kilometers of canals were built between 1960 and 1963 to reclaim 550,000 *mou* from a total of 800,000 *mou* previously inundated or waterlogged.[49] In the winter of 1963–1964, drainage projects were carried out throughout Ninghsia to expand the irrigated area by 260,000 *mou* and to improve productivity on 800,000 *mou*.[50] In the same period an inverted siphon was used to take excess water from the T'ang-lai and Han-yen canals under the main Yellow River channel to irrigate 40,000 *mou* on the east bank.[51] In the winter of 1964, 160,000 *mou* of new drainage projects were completed, with conditions on 170,000 *mou* of fields improved.[52]

The South Ninghsia Irrigation District is typical of medium-sized projects. Twenty-two reservoirs were planned in 1958, of which sixteen were completed by early 1960. The planned storage capacity was 820 million cubic meters and the planned irrigated area was 3.5 million *mou*, of which one million *mou* were under irrigation by 1960.[53]

Construction of the Chang-chia-wan Reservoir, largest on the Ch'ing-shui River in southern Ninghsia, illustrates the conditions under which these projects were built. The first stage of construction which included an earth fill dam 1,500 meters long and 25 meters high with a fifty-kilometer main canal, was completed within three months during the summer of 1959. There is no indication that state funds were invested, and construction was carried out by cooperating labor groups from various communes in T'ung-hsin Hsien.[54]

Elsewhere in the middle portion of the basin similar emphasis on medium- and small-scale projects is apparent. Two medium-sized projects have been reported in the Wei Valley of Shensi. The first one, completed in 1962, used a 103-kilometer canal to raise water, in staircase fashion, and received state funds for construction of the canal and pumping stations.[55] The second one, the "East Is Red" Irrigation System, finished in 1970, similarly raised water to irrigate the Wei-pei Plateau. In this project thirty pumping stations raise the water eighty-six meters in eight stages.[56] The relationship of these two undertakings to the various projects in the Kuan-chung Plain built or improved during the 1940s (Chapter 3) is not clear. Recent reports from China have not mentioned the earlier works by name.

In Kansu the best-known small project, the Bravery Canal, was built in Yung-ching County. A thirty-five-kilometer canal was constructed to irrigate thirty-five thousand *mou* and was hailed as a great achievement because of the difficulties overcome in constructing an aqueduct across a gorge of the Yellow River.[57] Like many similar projects, the Bravery Canal was widely publicized as a model for other local units to emulate in water management construction.

Electric pumping stations and pump wells have taken over much of the irrigation water supply work in the middle portion of the basin, as they have on the plain downstream. A typical report describes two stations completed near Wu-chung, Ninghsia, in 1965, for reclamation of ninety thousand *mou*. At the beginning of that same year there were fifty-seven such stations for irrigation in Ninghsia and fifteen for drainage.[58] In Yung-chi County, southwest Shansi, construction on a system of wells was started in 1953, but it was not completed until 1970, after greater importance had been attached to this type of undertaking.[59] In Shensi Province 2.7 million *mou*, or twenty percent of the total cultivated area, were irrigated by wells in 1965.[60]

To summarize events in Yellow River irrigation development during the last decade and a half: a flexible strategy has been implemented which combines historical and modern methods. The policy was instituted largely in reaction to failures in the attempts to construct enormous irrigation schemes during the 1950s. Labor mobilization remains a basic feature of irrigation development, as does local initiative and autonomy in project design, construction, and operation. Electric pumps and other features of industrial technology are the principal additions to historical labor-intensive methods. This combination of labor- and capital-intensive means gives each local unit a great degree of flexibility in solving its water management problems. If a unit is deficient in either capital or labor resources it can emphasize the other one; if a particular problem calls for more or less of one input relative to the other, the local unit has the possibility of adjusting its approach.

One ramification of this strategy is that it appears to have alleviated, temporarily at least, the earlier perceived need for transferring water from south China into the Yellow River basin. The development of numerous local storage projects and the opening of ground water resources through well irrigation appears to have altered the earlier conclusion that the Yellow River basin is deficient in total water resources. These two measures were mentioned by Soviet experts in 1959 as the only alternatives to interbasin transfers. Their implementation presumably is the reason that Chinese officials now state that the basin is not deficient in water resources (see Chapter 4).

Another feature of irrigation development which is suggested by comparisons of modern irrigation practices with historical methods is the role of reclamation projects in the Great Bend and Ninghsia. Historically, new and vigorous Chinese governments expanded their influence to the edges of cultivation. The main mechanism for this occupation of the frontier was agro-military colonization by garrisons which operated the vast irrigation schemes and defended the settlements in the area. Modern state farms do not represent exactly the same type of settlement, and the acquisition of industrial technology suggests that the modern Chinese occupation of the frontier will be more permanent than any historical precedent. Yet the state farms, employing soldiers and resettled peasants to reclaim and cultivate marginal lands, certainly are reminiscent of earlier frontier methods.

Reclamation by state farms was one of the earliest measures adopted by the Communist government in the vicinity of the Great Bend. One large project, reported in Inner Mongolia in 1956, involved nine thousand workers, including two thousand youths from Shansi, Shensi, and Hopei who were to join forces in the eastern part of the Great Bend.[61] A total of twenty-five state farms was reported in Ninghsia in 1964, including ten in the "Seventy-two Connected Lakes" region.[62]

Soil and water conservation is another feature of Yellow River management greatly affected by the changes in management strategy of the past fifteen years. The problems confronted in antierosion work were not the same as those in irrigation work, however. Not only were early mass mobilization campaigns insufficient to accomplish the desired goals, but the seriousness of the problem was admittedly miscalculated. It has proved to be a much more difficult problem to solve than it was considered to be during the 1950s. Three different types of measures have been instituted to augment the mass labor campaigns for erosion prevention work. These measures are increased scientific research, silt-settling basins on the lower course of the Yellow River, and reservoirs in the middle portion of the basin for impoundment of the silt.

Scientific research on erosion prevention measures was carried out from the earliest years of the Communist government, as a continuation of the pre-Communist beginnings in this field (Chapter 3).[63] The scope of investigation was greatly expanded during the 1950s as the technical aspects of soil and water conservation were developed and increasing numbers of technical workers were trained.[64] The new impetus given to Yellow River management after 1960 brought about renewed efforts in antierosion work, along with increased scientific experimentation and research on other aspects of the Yellow River problem. Soil and water conservation experimental stations were increased in all the provinces and autonomous regions of the middle portion of the basin.[65]

The new program was based on the use of certain antierosion and agricultural experiment stations that had been quite successful over a number of years and could be held up as models whose experiences could be emulated. Chief among the model stations were the Ch'iu-yuan-kou Station near Sui-te, Shensi, the Hsi-feng Station in eastern Kansu, and the Li-shih Station in Shansi.[66] Such model programs continue to play an important role in educating and mobilizing the masses for antierosion work. The most frequently mentioned types of work are check dams in gullies, the construction of field terraces, the improvement of cultivation methods, and the planting of tree, shrub, and grass covers.

Experiment stations for research in reclamation methods have played a similar role in expanding cultivation in marginal lands. Such stations at Chung-wei and P'ing-lo in Ninghsia and at Bayengol, Inner Mongolia, have carried out the most extensive programs of experimentation with cultivation of alkaline soils and desert control methods. The latter methods include not only planting shelter belts to impede the advancement of blowing sand, but also the diversion of Yellow River water to irrigate desert soils.[67] Silt-laden water is used to flood irrigate some fields, which brings a fertilizing layer of alluvial soil. Such fields, with up to fifteen centimeters of new soil, are used in Ninghsia to expand rice cultivation. This method of land enrichment reportedly was not practiced in the past when the degree of control over the river was not what it is now. The fragmentary nature of private land holdings made it impractical.[68]

Silt-settling basins and purposeful accumulation of silt in reservoirs are measures implemented in conjunction with the decision to reconstruct the San-men Gorge project so that silt can pass through its dam. These measures are necessary because erosion control alone will not solve the silt problem for some years to come.

On the lower course of the river, silt-laden water is pumped through or over the levees into large basins where the silt is deposited to strengthen the levees and to create fertile land for cultivation. The clear water then is used for irrigation or put back into the Yellow River to help scour and stabilize the main channel.

This program has been greatly expanded since the Cultural Revolution, and apparently is quite successful. Three hundred kilometers of dikes have been strengthened in this way, and eight million *mou* of farmland, approximately 15 percent of the total irrigated area in the basin, is resilted land. With deposition of three to five meters possible in the settling basins, a considerable potential exists for future disposition of the river's silt burden. Furthermore, the necessary pump stations or diversion works are relatively small-scale and can be undertaken by local units.[69]

Success with this method contributes to flood control by stabilizing and deepening the main channel, and it furthers irrigation and power generation by helping to reduce the silt load. Such successes certainly are a cause for optimism and are undoubtedly related to the recent resumption of official progress reports on major Yellow River projects.

There are disadvantages to relying on settling ponds rather than on erosion prevention, however. First, the continued silt burden greatly reduces potential for management in the important section of the river between the Loess Plateau and the floodplain. The T'ao-hua-yu project has been postponed, and the San-men Gorge project was greatly altered to allow silt to pass through the reservoir and power station. Without erosion control similar compromises would have to be made at projects from the Shensi-Shansi border stretch of the river to Hsiao-lang-ti, site of the highest dam and largest power station in the original plan. Hsiao-lang-ti and other projects in adjacent gorges would bolster the Chengchou-Loyang-Sian power grid, which is responsible for industrializing a large section of north China.

More basic perhaps is the question of how permanent a solution this measure actually represents. As basins near the levees fill with silt and water must be pumped increasing distances for settling, the difficulties will increase toward a point where this no longer will be practical. Erosion control seems, therefore, to be the unavoidable ultimate solution to Yellow River management. The current program does represent an excellent method for buying time while erosion control measures are perfected and more widely implemented on the Loess Plateau.

The use of reservoirs for holding silt along the main tributaries in the middle Yellow River basin is the least publicized measure taken to solve the silt problem. Very few details are known about the location or operation of these reservoirs, yet this method recently was described as more important than any measure taken for silt control on the main stem of the river.[70] Eighty-five medium-scale reservoirs (ten million to one hundred million cubic meters capacity) are reported to have been constructed on the main middle basin tributaries.[71]

Like the downstream settling basins, these reservoirs do not represent a permanent solution to the silt problem since they will fill up relatively rapidly and since problems may arise on how discharge will continue without eroding the accumulated silt. At the same time, however, these reservoirs offer a useful holding action against the silt until it can be stopped at its source—the slopes from which it is originally eroded.

The information available on major tributary projects is presented in Table 5-4, and general plans for development of the main tributaries are summarized in Appendix 4. It seems unlikely that these

Table 5-4. *Principal Projects on Yellow River Tributaries*

Basin	Project	Size	Benefits	Progress
Fen River	Ching-lo	ht 63 m lt 700 m sc 800 m cu m	gc 6,000 kw ia 1.5 mm	b 1958 pc May 1959
Wei River	Pao-chi Gorge		ia 1.8 mm	b 1959
Ching River	Ta-fo-szu	ht 62 m sc 150 m cu m	ia 3.3 mm gc 20,000 kw	b 1959 c 1960
T'ao River	Ku-ch'eng	ht 42 m lt 664 m sc 286 m cu m	gc 11,000 kw ia 1.5 mm	b June 1958 pc 1959
Yi/So. Lo River	Ku-hsien Tung-wan	sc 350 m cu m sc 675 m cu m	ia 4.5 mm gc 50,000 kw Flood prevention	b 1958

Abbreviations:

ht:	height.	mm:	million *mou.*
m:	meters.	m cu m:	million cubic meters.
lt:	length.	b:	construction begun.
sc:	storage capacity.	c:	construction completed.
gc:	generating capacity.	pc:	construction partially completed.
ia:	irrigated area.		

Sources: Various NCNA reports.

large-scale reservoirs would be allowed to fill with silt, because of the consequent elimination of their flood control, irrigation, and other benefits. Yet they certainly must be included in the eighty-five reservoirs referred to above. As with other features of the over-all Yellow River scheme, no definitive statements can be made on these projects until more information is made available.

From the discussion in this chapter it is apparent that the historical method of mobilizing large labor forces for the construction of water management projects remains a vital part of modern Yellow River management strategy. Labor mobilization is most important and has been most successful in the construction and the maintenance of flood prevention levees and detention basins. In irrigation development and soil conservation work mass labor campaigns were heavily relied upon during the decade of the 1950s, but have been modified considerably by new techniques during the past fifteen years. Those techniques combine modern technological advances and local initiative and self-reliance. The result is an approach to water resources problem-solving which allows a great amount of flexibility to the unit responsible. It also fosters experimentation with new management methods. In this context it is clear that a fixed approach to river management has by no means been determined. The strategies for Yellow River control are evolving and probably will remain so for some time.

Chapter 6
The New Strategy and
Its Implications

The features of Yellow River control discussed in preceding chapters suggest certain patterns of water management which are very significant in a comparative framework. The flexibility built into the river management strategy is one salient point, as are the social parameters of development and the resource management philosophy reflected in the scheme. Before these three points are discussed in detail, however, mention must be made of the larger context in which the river control scheme is being implemented. That context is the process of industrialization which is transforming the Chinese economy.

In the broadest perspective, the role of water management in the Yellow River basin actually is becoming less significant than it was in the past. Formerly maintenance of the irrigation systems on the Great Bend was a sine qua non for Chinese control of the inner Asian frontier between the Chinese and nomadic societies. With the irrigation works intact, the Chinese could control this transitional zone which was not perfectly suited either to their agricultural practices or to the pastoralism of the steppe nomads. Without the irrigation canals, Chinese claims to the area had no foundation in economic fact. The historical ebb and flow of armies and empires across the Great Bend area reflected the situation in which no group was able to establish a permanent hold on the region.

Lattimore referred to the end of these historical conditions, however, when he wrote: "It is the penetration of all Asia by the European and American industrialized order of society that is putting an end to the secular ebb and flow by making possible—indeed imperative—a new general integration."[1]

More than thirty years later the evidence of this integration is clear. In the first place, population has shown significant increases in all provinces of the middle Yellow River basin in the last two decades (Table 6-1). Much of the population growth has come from migration to the middle basin from the provinces of east China.[2]

More fundamental to regional integration is the development of industry in the middle basin. Pao-t'ou, Taiyuan, Sian, Lanchou, Sining, and several secondary cities have become regional centers for iron and steel production as bases for industrialization.[3] Other industries, such as petroleum refining in Yu-men and Lanchou, textile manufacturing in Sian and Hsien-yang, coal mining at Ta-tung and Shih-tsui-shan, to name a few, bring the area firmly into the fabric of the national economy. Regarding transportation, the T'ien-shui–Lanchou Railway and the Pao-T'ou-Lanchou Railway were completed in 1951 and 1958, respectively, and a line had been extended to fifty kilometers west of Sining by 1960. The development of highways and other facets of the spatial economy further serves to integrate the middle basin into the national economy.[4]

These facts emphasize that water management is only a part of modern Chinese development of the middle Yellow River basin. Whereas historically water management was the single key eco-

Table 6-1. *Population Growth in the Middle Yellow River Basin*
(Millions of People)

Province or Autonomous Region	1957 Registration	1965 Estimate
Chinghai	2.05	2.4
Inner Mongolian Autonomous Region	9.2	13.4
Ninghsia Hui Autonomous Region	1.81	2.0
Shansi	15.96	18.0
Shensi	18.13	21.0

City	1953	1970
Lanchou	0.397	1.45
Taiyuan	0.721	1.35
Sian	0.787	1.31
Pao-t'ou	0.149	0.92
Sining	0.094	0.50

Source: Chen Cheng-siang, "Population Growth and Urbanization in China, 1953–1970," *Geographical Review*, Vol. 63, No. 1 (January 1973), pp. 534–548.

nomic factor that allowed the Chinese military/agricultural colonies occasionally to control the area, now it constitutes only one element in Chinese economic use of the area. It certainly is a basic element, helping to supply the agricultural, power, and transportation needs of the region; but it is by no means the sole factor in modern Chinese obliteration of the historical frontier.

On the floodplain, similarly, the role of water management is changing. No longer is the posture of the Chinese people toward the river basically a defensive one. Flood prevention is losing its primacy among water management tasks, while irrigation, power generation, and navigation assume more significance. Because modern Yellow River management is developing in this context, the possibility of applying its salient features to other industrializing countries is timely indeed.

The Sociological Fix

The labor-intensive element in Chinese water management practices is the element that contrasts most sharply with Western practices. The history of water control in the Western world, from Egyptian dams and Roman aqueducts to modern sprinkler irrigation systems and desalinization plants, is viewed as a series of technological triumphs. Even in premodern times when major dams and waterways required massive amounts of labor, the attention of the West was always drawn to engineering practices.

In China, on the other hand, technology was only part of the overall science of water management, no more important historically than the philosophy of management or the mobilization of labor for management works.[5] During the period of foreign involvement in China, from the middle of the nineteenth century until 1960, the technological aspects of management received greater emphasis; but the recent strategy of relying on the masses in the Yellow River basin and throughout China can be interpreted as a reversion to capitalizing on the major resource of surplus agricultural labor, albeit with modified but more effective organizational forms.[6]

While reliance on technoolgy has prompted the economic development characteristic of Western civilization in the last several centuries, it also has been a root cause of that development's negative environmental impacts. Other aspects of Western society have been criticized as contributing to environmental degradation,[7] but many writers from diverse fields point to our use of technology as the fundamental cause of the environmental crises.[8] Critics of over-reliance on technology point out that the real danger lies in thinking all our resource management problems can be solved by more and more machines. This narrow "technological fix" attitude is widely refuted, and numerous examples have been cited to illustrate its bankruptcy.[9]

The clearest analysis of exactly how our reliance on technology has led to the current ecological crisis is provided by Barry Commoner. He points out that our use of technology is reductionist in nature, based on the premise that effective understanding of a real, complex system can be achieved by investigating the properties of its isolated parts.[10] Such an application of technology often solves a specific problem, while aggravating related problems in the same systematic framework.

Recent Nile River development is a widely known example of this phenomenon in water management.[11] Construction of the Aswan Dam provided clear benefits in the areas of flood control, irrigation, and power generation. The impact of the dam on hydrologic, biotic, and human health features of the basin was catastrophic, however. Such features, which obviously are related in a single ecological complex, simply were not accurately considered in the implementation of the partial technological solution.

No information is available on the impact of major Yellow River dams on aquatic plants and animals in the basin. There is evidence of a reductionist tendency in the application of technology to Yellow River management in the 1950s, however. The criticism leveled against those who emphasized only irrigation and ignored related drainage probems (quoted in Chapter 5) shows how much easier it is to approach an isolated problem with technological solutions than to address a complex of related problems. Similarly, the conflicts between agricultural, conservation, and forestry practices on the Loess Plateau appear to illustrate a fragmentation of the problem rather than a holistic approach. Perhaps the most obvious example is the enthusiasm with which the dam at San-men Gorge was undertaken initially, in spite of clear statements of the problems its construction would raise.[12]

The shift toward more reliance on labor mobilization and related strategies since 1960 appears to help circumvent the reductionist approach. This is accomplished by emphasizing social rather than technical means. New technology certainly is utilized, as the growing number of siphons and pumps illustrates, but primary reliance is on organizational forms rather than on technological forms. Thus it is fair to state that this strategy is characterized by a "sociological fix" rather than by a "technological fix."

In this strategy the motivation of the masses to undertake labor projects is the major task of leaders and functionaries, just as the major task in Western society is the effective application of machine technology. The development and popularization of new organizational forms corresponds to the invention and dissemination of technological advances. It is not surprising, therefore, that so much attention has been given in recent years to the techniques of mass labor mobilization for water conservancy work.

In mobilizing the masses for labor projects four major guidelines are usually followed: (1) conducting ideological education, (2) insuring the belief that the work is suited to the interests of the masses, (3) implementing rational construction burdens, and (4) showing concern for the livelihood of the masses and safeguarding their strength.[13] The first of these guidelines, ideological education, encourages the laborers to consider the fruits of their work in patriotic terms and in international contexts, as well as in relation to local and regional development. The second guideline is aimed at developing more "vigorous and healthy" mass movements where the projects benefit the people who are called upon to provide labor. The rational burden concept means that those who benefit most from a given project must contribute most toward its construction, and vice versa. The fourth guideline, showing concern for the masses and safeguarding their strength, is an exhortation to leaders for rational living, work, rest, and health conditions for laborers. If these guidelines are followed, it is said, mass movements for water management construction will be successful. Other policies designed to "arouse the enthusiasm of the masses" include making grain on newly reclaimed or terraced fields exempt from taxes for three to five years and giving "honorary awards and material incentives" to those who achieve distinguished results in water and soil conservation work.[14]

The extent to which material incentives have been used is difficult to assess, though such measures have been broadly curtailed since the Cultural Revolution. The honorary awards technique is widespread and very successful, however. It is illustrated by the use of labor models in water management work. Successful projects in key areas, such as the well-known Tachai Brigade terraced fields in Shansi, and the Red Flag Canal in Lin-hsien, Honan, are widely publicized.

A less well known example is the case of the Meng River basin development effort on a small tributary of the Yellow River in western Honan. This project received a banner at the Second National Water and Soil Conservation Conference in December, 1957, for accomplishment in its development plan. The honor was publicized nationally and the project became a model for small basin development throughout China.[15]

By appealing to the pride of accomplishment, leaders hope to inspire similar success in other areas through emulation of these models. Beyond the pride inspired and the motivation for greater effort induced by such campaigns, model emulation has deep roots in Chinese society as a teaching vehicle. For centuries correct behavior was taught by holding up model individuals who displayed the desired behavior for the people to learn from.[16] In the case of labor mobilization the effect is to disseminate and make popular new water management techniques.

Three major principles underlie the "sociological fix" in Chinese water management. The first is a belief in collective strength. Wang Hua-yun, director of the YRWCC, stated in a 1957 speech that large-scale terracing and other measures were impossible when individual peasants were concerned only with their own small private plots.[17] Numerous subsequent statements have emphasized the point that collective strength is the basis of successful water management. Production teams are encouraged to plan and undertake their own projects, and to cooperate with each other on projects requiring more labor. Brigades and communes similarly work together on larger projects that benefit more people.[18]

The second principle is known as "taking agriculture as the point of departure" for water conservancy projects. This approach appears to de-emphasize long-term projects and planning and instead to concentrate on works which "show benefits in the same year." This may indeed foster shortsighted development goals, yet agencies at all levels do develop long-term plans as well as short-term plans.[19] More important to sustaining the motivation function of the policy is the immediate return for labor invested in water management construction.

The third principle is the most significant in a comparative framework. The "sociological fix" promotes an integrated attack on all related water management problems by the local unit. This strategy grew directly from wastes incurred by the partial completion of large specialized projects in the 1950s. The construction of auxiliary projects to render earlier unfinished projects more useful is an integral part of recent management strategy.[20] This also is articulated as "multiple purpose control in light of local conditions" and "comprehensive treatment and overall planning" (as discussed in Chapter 5). A local group attempts to solve all of the related problems in its area, rather than concentrating on a single large problem that encompasses a large area.[21] If this strategy proves successful in accomplishing such ends over a long period of time, it may have much to offer to Westerners who are becoming increasingly disenchanted with the "technological fix" approach.

Transforming Nature

Foreigners interested in modern China have been struck by the contrast between the contemporary rhetoric used to describe the management of China's natural resources and the traditional values ascribed to Chinese philosophy. Constant references in recent Chinese publications to "overcoming nature" or (as regards the Yellow River) to "conquering the river" lead to the conclusion that the general attitude toward the natural world has changed radically from earlier Taoist, Buddhist, and even Confucianist values and precepts. Rhoads Murphey has suggested that the "revolution in the conception of man's relationship to his physical environment" is among the most radical effects of the Communists' assumption of power.[22]

In the specific area of river management we may note that the change did not occur suddenly in 1949. Traditionally Yellow River managers relied on the principle of proceeding "according to the nature of the water" or "using nature to control nature."[23] This was replaced by the approach of early-twentieth-century Chinese engineers to "use science to control the river."[24] This principle in turn has been replaced by Communist exhortations to effect river control through overcoming nature.

Problems arise when the Communist rhetoric concerning the "war against the environment" and the "war against nature" is taken at face value, however. The greatest danger is that the Western reader is tempted to transfer the struggle defined in these terms into the more familiar context of our own assault on nature, so vividly described in environmental literature.[25]

It is evident that Chinese publications also speak of "restructuring nature," "reforming nature," and "changing nature" more often than they speak of vanquishing nature.[26] The significance of this range of rhetoric can only be understood within the context in which it is written.

The large majority of Chinese press materials from which such references come, and through which Western social scientists study modern China, are articles and editorials with a very specific purpose. These materials are written by party, government, scientific, or technical leaders. Their purpose is to report conditions and to outline programs for reaching the goals of such leaders—national development. These are not philosophic statements, but rather reports and suggestions on the development of concrete programs.

Regardless of whether the subject is irrigation, water and soil conservation, afforestation, or any other enterprise, such writings display a very consistent pattern. The goal of the work is described as socialist construction, or economic development; the specific problem (irrigation, conservation, forestry, etc.) is described as a task in the struggle with nature to attain that goal; and the main means to accomplish the task is the motivation of the people toward the desired ends. Statements on the struggle with nature consequently should be taken within the context of education and mobilization of the masses. They represent morale-building rhetoric rather than profound philosophical expressions.

Understanding these statements in their context helps to account for the range of rhetorical expressions employed. The most vehement language is used for the most pressing tasks. A river with flood waters threatening, or a serious drought, is to be "conquered." A longer-term, broader-scale conservation or development program will "transform" nature, not conquer it.[27]

Some articles state explicitly that social and political conditions (for mass mobilization) are more important than the relations between society and nature. An article entitled "Planting of Trees Must Be Preceded by Nurturing of People" states that "to develop forestry production with plentiful, quick, and good economical results, we must believe in the masses, rely on them, and follow the mass line."[28] It states further that "after the trees are planted it is all the more necessary for everyone to show care and love for them, only in this way can they survive in large numbers and grow well." The goal of economic development, the task of afforestation, and the emphasis on mass mobilization are all very clear. Equally clear is the absence of any sense of the wanton assault on nature which often has attended Western economic development, or even superseded development for various reasons.[29] The attitude reflected by this behavior toward the natural environment is a sober, utilitarian approach to the economic exploitation of resources.

Similar relationships are reflected in another work, entitled "To Reform Nature It Is Necessary to Reform Thought First."[30] Here "thought" refers to the correct methods of education for labor mobilization. Education is required because incorrect political thinking—"rightist conservation"—had previously hampered socialist construction.

This article foreshadows the political context into which resource management philosophy was placed during the Cultural Revolution of the late 1960s. The Cultural Revolution is reported as a defeat for anticonservation forces. Prior to that time "Liu Shao-ch'i and his agents" and "class enemies and capitalist forces" are said to have wantonly and indiscriminately felled trees, and to have advocated "leveling of forests to reclaim land." Their erroneous theory that "large populations destroy forests" was based on the division of forest exploitation rights down to the level of households. Their defeat resulted in the protection of forests from individual exploitation through state and collective ownership.[31] Once again, such reports are difficult to reconcile with what is meant in Western society by the term "war against nature."

A few reports from the Yellow River basin even appear to suggest an attitude more in harmony with nature than in conflict with it. Reclamation workers on the Great Bend are said to have "used nature to control nature" in planting low shrubs in rows four meters apart in the desert and allowing sand to blow against them to form natural canal dikes.[32] Siphon irrigation workers in Shantung overcame failures not by more aggressive assaults on nature, but by "mastering the law of mutual relationships among water, mud, sand, land, and crops" and by analyzing human failures to turn harm into benefit.[33] The basic precept which this latter group followed was "seeing the benefit in harm and preventing harm in benefit (*hai chung chien li, li chung Fang hai*)." This precept is closer to a Taoist view of natural phenomena than a materialist view. More recently the use of the river's silt load to strengthen the flood prevention levees in silt-settling ponds was described as "using the river to control the river."[34] This is precisely the principle which guided the famous Yellow River managers in traditional China.

Taken in context this range of rhetoric regarding the struggle for river control and general nature transformation certainly does not imply a war on nature simply for the sake of defeating nature. Even the strongest statement by Mao Tse-tung on the transformation of nature does not imply such a goal: "For the purpose of attaining freedom in the world of nature, man must use natural science to understand, conquer, and change nature and thus attain freedom from nature."[35]

Such a utilitarian rationale for the transformation of nature by human civilization is reminiscent of the ancient labors of Yi and Yu. These legendary heroes are credited with first having made the land habitable for Chinese civilization by river regulation, swamp drainage, and jungle clearance.[36] In the modern era, of course, such accomplishments are not credited to legendary heroes but to the laboring masses. One task which was never accomplished by the ancients, that of making the Yellow River run clear, is at least closer to the grasp of modern managers.

Evaluation of Chinese Water Management Strategy

The unique water management strategy in the Yellow River basin has evolved as a combination of traditional management practices, influences from European-based industrial society, and new technical and organizational features from modern Chinese society. The importance of various elements in the strategy and the degree of success of each have altered in the past twenty-five years as overall Chinese development strategy has changed.

Flood control has been the fundamental element of the strategy, and must be considered the most successful. It received highest priority during the early years of the new regime and although it has received a smaller proportion of the total management effort since that time, it remains the basic feature without which the other facets of development could not be carried out.

It has also been the element most dependent on traditional labor mobilization methods. Mobilization of hundreds of thousands of peasants in the agricultural off-season for earth moving to build or strengthen levees, to construct flood detention basins and carry out dredging work, is the closest continuation of age-old practices.

After the first decade of such development the task became one of maintenance, since basic construction was completed. With the opening of key modern dams upstream, which help to hold back the highest flood crests, the flood hazard has been even further neutralized.

Flood control has certainly accomplished its original tasks more fully than other aspects of the basin-wide scheme. A total of one thousand kilometers of flood prevention levees have been rehabilitated by traditional means along the lower course of the river, involving 390 million cubic meters of earth and stone work in twenty-five years. Three major flood detention basins have been constructed and three hundred kilometers of levees have been strengthened by deposition of silt in settling basins. In spite of high flood crests in at least two years, 1954 and 1958, major flood damage has been avoided since the early 1950s.

Responsibility for construction of flood control works has remained in the hands of provincial River Affairs Bureaus (Ho-wu-chu), premodern units that have survived to carry out levee construction and maintenance work. They receive coordination and advice on technical and planning matters however, from the YRWCC, the modern basin-wide development agency. And they must coordinate with county and commune governments in the organization of labor forces.

Irrigation is the next most basic and next most successful aspect of the scheme. A total area of 48 million *mou*, nearly one-sixth of the cultivated area in the basin, was under irrigation by the end of 1974, representing a fourfold increase over the irrigated area of 1949, and surpassing the target of 42 million irrigated *mou* set for the first phase of development. That this target originally was to have been reached by 1967 shows how irrigation development has not met early expectations; yet it is further ahead of revised plans than are some aspects of the scheme. One-sixth of the total 48 million *mou* of irrigated land are newly silted, and an additional 1.7 million *mou* are improved sandy or alkaline soil.

Irrigation is also the aspect of the scheme which combines traditional and modern elements most successfully. In the 1950s irrigation development was attempted on a large scale through mass labor mobilization projects that met with limited success. Since 1962, and increasingly since the Cultural Revolution, development has been characterized by the integration of labor-intensive canal construction work and capital investment in pumps and other equipment in a single project. More than eighty pumping stations and siphons, along with sixty-five concrete sluice gates, now divert water on the lower reaches in place of the dozen huge, labor-intensive diversion schemes planned during the 1950s. Pump wells developed at the local level account for 15 percent of irrigation in the Yellow River basin.

Likewise, irrigation development illustrates the multilevel structure of organizational units in the modern scheme. National water management directives, such as those emphasizing small-scale projects beginning in 1957 and 1958, are issued by the central government, and are implemented at the provincial, municipal, county, or commune level (and even lower) in accordance with plans for projects of a certain size. The locus for responsibility in irrigation development has shifted downward in this hierarchy from provincial to county and commune levels, as smaller-scale projects have been emphasized over large projects, contributing to what is in effect a decentralization of the irrigation development effort.

Development of hydroelectric power generation has proceeded more slowly, due in part to problems encountered in construction and in part to power generation being given a lower priority in revised plans. It did not reach the original 1967 goal of 2.3 million kilowatts installed capacity until early in 1975. The Liu-chia Gorge and Yen-k'uo Gorge projects are the cornerstones of power generation, having realized the originally planned installed capacities of approximately one million kilowatts and half a million kilowatts respectively. A generating capacity of 260,000 kilowatts has also been completed at Ch'ing-t'ung Gorge as planned. The San-men Gorge power plant has been completed with only one-fifth of the originally planned 1-million-kilowatt capacity, and construction apparently has not begun on a 50,000-kilowatt power plant at T'ao-hua-yu. A major reason for these modifications and delays is the continued difficulty in solving the silt problem, which remains the most serious obstacle to overall Yellow River development.

While this aspect of development has necessitated the greatest concentration of capital-intensive technology and the greatest dependence on foreign powers, the projects involved have not excluded the use of traditional methods as well. At Ch'ing-t'ung Gorge, for example, the concrete dam structure (697 meters long and 42 meters high) and the power plant (260,000 kilowatt installed capacity) required modern technology and considerable capital investment to build. However, 183 kilometers of trunk canals and the vast network of irrigation and drainage ditches in the 7,000-square-kilometer irrigation district were renovated or constructed at the same time with integrated labor-intensive methods.

The priority for this and larger power generation projects undertaken by the central government has been lowered relative to irrigation and silt control since the first Five Year Plan, however. The role that such projects were to play in supplying power to modern industrial plants has been eclipsed by emphasis on rural electrification, for which small units operated by county or commune administration have come to assume 20 to 30 percent of the power generation burden.

The river's silt load remains, as it has for centuries, the root of all other problems of river management in the basin. Silt control must be treated as a separate purpose of development with its own priority. The experience of the past twenty-five years shows that the silt problem cannot be solved in a short time, either with the application of modern techniques or with a combination of modern and traditional techniques.

The ambitious mobilization of tens of thousands of antierosion workers during the late 1950s provided very limited and partial solutions to the problem, and has been replaced by a three-sided approach: scientific work is carried out through conservation experiment stations, which have been expanded on the Loess Plateau; technological development is represented by the use of pumps and siphons at silt-settling ponds on the lower course; and intensive labor has been used to build earth fill dams for silt detention reservoirs on the Yellow River tributaries. These endeavors have produced incremental improvements, but have shown that a final solution to the problem will be a long time in coming.

The admitted miscalculation on the amount of silt which would accumulate in the San-men Gorge Reservoir not only forced modification of its dam structure and power plant, but also prompted revision of antisilt measures. Currently reservoirs on the main tributaries and settling basins along the river's lower course are being used to accumulate silt, while erosion prevention measures are improved and more widely implemented on the Loess Plateau. In early 1975, some eighty-five large- and medium-scale reservoirs (at least ten million cubic meters capacity each) were serving this function on the middle basin tributaries, and eight million *mou* had been resilted through settling ponds on the lower course of the river. These measures represent a useful holding action against the silt problem, but the unavoidable ultimate solution lies in stopping silt at its source through erosion prevention measures.

Three other goals of development—navigation, municipal and industrial water supply, and interbasin transfer—have a relatively low priority at the present time. Navigation is developed to some degree in the middle portion of the basin, but must wait for improved channel stabilization in the lower course for further significant development. Very little information is available on water supply, although it is known that Yellow River water is used for parts of the municipal and industrial requirements of such major cities as Chengchou, Kaifeng, and Tsinan. The current status of plans begun during the 1950s for interbasin transfer is also unclear, except for a very few scattered reports of small transfers.

The unique water management strategy developed in the Yellow River basin, blending traditional and modern methods, encouraging small- and large-scale projects, is intimately related to administrative structure. The structure of administrative agencies is also unique among major river basin projects of the world. Basin-wide planning is the responsibility of the Yellow River Planning Commission, a staff office of the State Council, and therefore a unit of the central government. Administration and other functions are in the hands of the Yellow River Water Conservancy Commission (YRWCC), which is primarily an advisory and consultative unit. All facets of survey and research work in the basin are within its sphere, but are not its exclusive domain. The YRWCC provides data and planning alternatives to the large dam project bureaus, for example, but other work, such as hydraulic model test experiments, is conducted at a variety of laboratories in addition to those of the commission itself.

Among its functions, which grew from its inception in the 1930s, is advising provincial bureaus on technical and planning aspects of flood prevention levees. However these bureaus answer to their provincial governments on construction matters. Similarly, for irrigation diversions or soil and water conservation experiments undertaken by counties and lower units, the basin commission coordinates efforts in the overall scheme; but it is not a strong, autonomous basin management unit with significant decision-making authority of its own.

The planning, design, construction, and operation of specific projects takes place at four different levels, depending upon the size of the project. The general principle is that responsibility for a project which affects two or more units is taken by the unit of higher rank. A corollary is that each unit which benefits from a project contributes labor and investment in proportion to its share of the benefit. Thus the central government, through project bureaus of the Ministry of Water Conservancy, takes responsibility for major dams and power station projects, and labor to supplement state investment funds is contributed by all provinces that will benefit. Provincial governments are responsible for irrigation projects that affect more than one county and/or municipality. County (municipal) governments usually undertake the diversions or reservoirs which affect more than one commune.

At the local level, commune or production brigade units plan, construct, and operate numerous projects of all types. These include small- or medium-sized irrigation diversions and reservoirs, pump wells, reclamation projects, and electrification based on small power generation stations. The principle of self-reliance encourages the local unit to undertake any project of which it is capable.

The efficiency of this structure, embedded as it is in the administrative framework that has evolved from agrarian Chinese society, lies in its ability to mobilize all the resources available—human and material—for water management work, while at the same time making as widely available as possible the limited scientific and technical resources of the developing economy. In this and in encouraging local innovation in construction and technical improvement, it has much to offer other nations, developing and developed alike.

Comparison of Chinese Water Management Strategy with That of Other Developing Countries

When the Yellow River strategy is compared to development programs in other major river basins with comparable management problems, its unique features come into sharper focus. From a broad historical perspective it might be expected that development of the Indus, Nile, and Tigris-Euphrates systems (where foreign engineers became involved in the 1820s, 1850s, and early twentieth century, respectively, and where recent political independence has been followed by a desire for modernization) would display similar patterns in the evolution of river management strategy.

Yet in each of the other three basins traditional methods have been wholly subordinated in recent decades to reliance on large capital-intensive projects. Egypt, Pakistan, and Iraq have become dependent on economic and technical aid from an industrializing country or set of countries for river management. This is the road from which China turned away, after the first Five Year Plan, in placing renewed emphasis on traditional methods in combination with modern ones to fashion river management strategy.

The basic reason lies in the differential impact of foreign techniques on indigenous methods during the period of foreign involvement in each country. Just as China resisted full colonial domination of its economic structure by foreign powers, it was able to avoid becoming dependent on foreign river managers while adopting the useful aspects of European hydraulic engineering.

As we have seen, the century preceding 1949 was a period when the groundwork was begun for modern development, but when little actual construction work was carried out. Foreigners studied the Yellow River problem, began designs for modern development, and contributed considerably to the education of Chinese engineers. Actual construction by foreigners was insignificant until the 1930s, however, and even then it involved local projects that provided local benefits for flood or famine relief. Foreign personnel and agencies were employed as consultants by the Chinese government. Their work was carried on as an adjunct to continuing traditional efforts—to the extent that the latter could be mounted by the Republican government.

Foreign involvement in the other basins, however, was of an entirely different nature. European engineers carried out basic construction projects for improvement of the Nile system, the explicit purpose of which was irrigation of long staple cotton for export in the international economy. During a century of such construction, traditional methods were dominated by foreign methods, were relegated to an inferior position, and stagnated as a result. It is not surprising that traditional methods had little to contribute in the mid-twentieth century and that a conditioned reliance on the capital-intensive program of industrialized countries was seen as the answer to river management problems, even by nationalistic leaders after political independence.[37]

Similarly, the nineteenth-century government of India undertook development of the Indus works for the economic and administrative benefits it offered to colonial interests. After a century of such development, not only had technical reliance on foreign methods become overwhelming, but the attendant organizational structure of development and the philosophical predispositions of Western science had divorced modern Indian river management from any elements of traditional society that might have served the development effort.[38] The entanglement over how partition of India and Pakistan should affect management of the Indus system, like the problem of the reservoir behind the high dam at Aswan reaching into the Sudan, was a further element requiring international arbitration and was without counterpart in the Yellow River basin.

River management in contemporary Iraq also emphasizes large flood prevention, irrigation, and power generation projects relying on technical aid and credits from Socialist bloc countries. It grew from a similar, though less lengthy, pattern of foreign renovation of historical projects which had suffered under the Turkish government. With revenues from petroleum to help finance construction, Iraq can more comfortably afford to import technologies than can the more populous, agrarian-based economies.[39] Yet the retention of traditional methods might have fostered a more even integration of modern river development projects into the rural economy.

In China alone among the "hydraulic civilizations" of the Old World have traditional practices survived in sufficient strength to affect the transition to modern river management. The transfer or emulation of specific elements of Yellow River management strategy in another country where old methods have been usurped by foreign methods, therefore, would be most difficult. And even more, an agrarian society that never had developed the technological, organizational, and motivational apparatus of traditional river management construction could not hope to initiate them under modern conditions where the pressure to industrialize pulls in an opposite direction—the direction of breaking down the fabric of traditional institutions.

Yet if specific elements cannot be transferred, the attitudes and principles which underlie Yellow River management should be understood and applied where possible. A willingness to retain and utilize any strengths that do remain from traditional culture is the primary one. Other countries might not have China's history of river management, but any indigenous strengths which do remain and which might be useful should not be cast off too readily in a rush to modernize or westernize. Any benefits of intermediate technologies that grow from indigenous roots should be considered in the spirit China has mobilized in Yellow River management.

Equally important is the Chinese example of finding organizational solutions where technological solutions are too costly or difficult to integrate. Yellow River management illustrates the main counterpoise, as developed in China, to the dominant Western belief that if a problem exists, a machine can be made to solve it. Few countries could adopt China's mass labor mobilization techniques, yet few are bereft of institutions that could be adapted to help solve indigenous problems in place of imported technological solutions or in concert with them.

Having surveyed the origins, strengths, and weaknesses of the Yellow River development strategy, we must not lose sight of the fact that it is a constantly changing adaptation by society to the natural environment. It has changed markedly since the Communist government arrived, and it will continue to change as society industrializes. Traditional methods are relied upon much less now than they were two decades ago. With increasing pressure from population and industry on the resources of the Yellow River basin, the new efficiency of machinery will demand further adaptations of the old efficiency of human organization.

The millions of tons of silt entering daily into the waters of the Yellow River systems have been decreased—if only by a fraction—in the past two decades. On the plain which the river built through deposition of the silt—and through frequent changes of its course —are rooted the communes which implement the new strategy to make the river run clear. The significance of the strategy is not a scientific measurements of the decrease in silt load, but the fact that it grew in large part from within the basin itself.

Appendix 1

Bibliography of Early Articles on the Yellow River Problem

(With comments by John R. Freeman; see Chapter 3, note 2)

Bickmore, A. S. "Some Remarks on the Recent Geological Changes in China and Japan." *Silliman's Journal*, 2d Ser., Vol. 45 (1869).

Elias, Ney, "Notes of a Journey to the New Course of the Yellow River in 1868." In *Proceedings of the Royal Geographical Society*, Vol. 40 (1870).

Fijnje van Salverda, J. G. W. *Memorandum Relative to the Improvement of the Huang Ho, or Yellow River, in China; Containing an Annex by His Deputies, Captain F. G. van Shermbeck and Mr. Visser*. Translated from the Dutch by W. P. Dickinson. The Hague, 1891.

Gandar, Le P. Domin. "Le Canal imperial: Etude historique et descriptif." *Varietes Sinologiques*, Vol. 4 (1894). ("A very valuable paper.")

Gordon-Cumming, Constance F. "The Yellow River." *Littell's Living Age*, Vol. 177 (1888), p. 99.

Jameson, Chas. D. "Flood Control in China." Paper No. 31, International Engineering Congress, San Francisco, 1915.

Lamprey, Dr. "Notes on the Geology of the Great Plain." *Royal Asiatic Society*, North China Branch, Vol. 2 (1865), p. 1.

Morrison, G. James. *Engineering*, Vol. 55. London, 1893.

———. "Journeys in the Interior of China—The Grand Canal and Yellow River." In *Proceedings of the Royal Geographical Society*, 1880, pp. 145 ff. ("Excellent map, figures representing depth of Yellow River in May, 1878.")

———. *Proceedings of the Royal Geographical Society*, 1879, p. 127.

Oxenham, D. L. "Overflow of the Yellow River." (British) Consul's Report for 1887. "Suggests reafforest the hills of Shansi and

Shensi, which are at present barren of wood, p. 535. Also cultivation of barren country at northern bend to prevent washing away of sand.")

Van Shermbeck, Captain F. G., and Mr. Visser. "Memorandum on the Hwang-ho." Printed for private distribution. ("The first part is by Fijnje, and is followed by a report from the other two engineers named.")

Appendix 2

Soil Conservation Priorities, 1946

(From W. H. Huang, *Soil and Water Conservation of the Yellow River Basin*, pp. 45–46)

Based upon the above understandings together with the data presented in this chapter and a study of Dr. Lowdermilk's authoritative recommendations a tentative scheme of soil and water conservation of the Yellow River watershed is outlined as follows (items are arranged in order of priority):

(1) Soil erosion and land use capability survey be carried out in the loessial plateau of Shensi, Kansu, and Shansi.

(2) Establish an extensive network of rainfall and discharge stations in the loessial plateau. Most of them should be equipped with automatic recording instruments.

(3) Establish at least one or two experimental stations in each of the following valleys (arranged in order of importance):

    (a) Wu-ting Ho,   (f) Tao Ho,
    (b) Upper Lo Ho,   (g) Fen Ho,
    (c) Nulu Ho,   (h) Lower Wei Ho,
    (d) Upper Ching Ho,   (i) Huang Ho,
    (e) Middle Wei Ho,   (j) Tsu-li Ho.

in order to study

    (a) relation between rainfall, runoff and soil erosion,
    (b) infiltration,
    (c) canopy interception,
    (d) suitable species of grass, leguminous herbs and shrubs and trees for revegetation for gullies and slopes,

(4) Carry out experiments to test materials suitable or most available and economical in the loess region for construction of bank protection works, silt control, and soil saving dams and their appertinent works.

(5) Construction of trial silt control and soil saving dams.

(6) Carry out reservoir sedimentation and bank protection studies both in the field and laboratory.

(7) Establish nurseries for growing of seeds of suitable species in great volume.

(8) Construction of silt control dams on the small tributaries and soil saving dams in the large gullies in the following valleys (arranged in order of relative importance):

(a) River basins along the Shensi and Shansi border,

(b) Upper Lo Ho,

(c) Upper Wei Ho,

(d) Upper Ching Ho,

(e) Shensi Plain,

(f) Tao Ho,

(g) Fen Ho,

(h) Huang-shui,

(i) Others.

(9) Extensive plantation of trees and shrubs on the river banks and in the gullies. Extensive construction of bank protection works along the rivers and checks, and check dams in the gullies.

(10) Revegetation of slopes.

(11) Retirement to pasture or woodland of steep, severely eroded land and land otherwise unsuitable for cultivation; rearrangement of fields to permit utilization of the land most suitable for crops; good soil management practices including rotation of crops, growing of legumes, etc., stripcropping and contour tillage; the use of terraces in suitable locations; and good management of pasture and woodlands.

Appendix 3
Critique of the Japanese Plan for
Developing the Yellow River

(From Eugene Reybold, James P. Growden, and John L. Savage, *Preliminary Report on the Yellow River Project*, p. 24)

A summary of the Japanese Plan for developing the Yellow River has been studied. It is a comprehensive plan covering the area from Pao-t'ou to the sea. It contains many excellent proposals and shows evidence of careful, intelligent study. It has the following defects:

(1) In the delta area it utilizes a very large amount of good arable land for flood and silt storage—thus reducing its productivity and value.

(2) The San-men flood control and power project floods a large area of arable land. It is not feasible to desilt the San-men reservoir and its useful life would be short.

(3) In the reach Pao-t'ou to Lung-men it provides eleven dams for power development. The reservoirs are small and would have only a short life.

(4) It offers no permanent solution to the problem of silt control. Hence the whole scheme of development would progressively become less efficient. All of the undesirable features of the Japanese Plan can be avoided and all of its good features can be attained.

Appendix 4
Tributary Development Plans, 1960
(Sources: NCNA reports and *Ajia no Yume*)

Ching-lo Project. The major feature of this project, located north-west of Taiyuan, is an earthfill dam faced with 4 million cubic meters of stone work. In addition to supplying 150–200 million cubic meters of water to industry at Taiyuan annually, 20,000–25,000 tons of fish also are raised for local consumption.

T'ao River Diversion Project. This project was tailored to serve the four million inhabitants of the arid plateau of southeastern Kansu. The 1,400-kilometer irrigation canal, which starts at Lin-t'ao Hsien, is navigable to hundred-ton vessels and has one hundred medium- and small-scale power plants along its course.

T'ao River. Two major dams were planned to regulate flow, irri-gate twenty million *mou*, and provide water for the T'ao River Di-version Project at Ku-ch'eng.

Huang-shui–Ta-t'ung System. Two major structures will be used to divert fifty cubic meters per second from the Ta-t'ung to the upper Huang-shui to provide water for irrigation, afforestation, industry, and power generation in the Sining Valley.

Ch'ing-shui River. Two major structures were planned to irrigate 32,000 *mou* of arid land.

Ta-hei System. Two dams will be built to solve the flood prob-lem and irrigate 720,000 *mou*.

Hung-shui River. Two dams will be built to protect and irrigate 150,000 *mou*.

Wu-ting River System. Eight major dams were planned for the Wu-ting and its tributaries to irrigate 2.64 million *mou* of arid and semiarid highlands and for power generation. Two of these reser-voirs had been completed by 1959.

Fen River. Two large dams on the Fen will provide multiple benefits to Taiyuan, and three dams on tributaries will irrigate nearly 1 million *mou*.

Northern Lo River. Three major dams were planned to irrigate 2.3 million *mou.*

Wei Basin. A total of thirty-one medium-sized dams will be built on the upper Wei and major tributaries principally for flood prevention and irrigation purposes, while forty-one relatively low dams will be built in the lower reaches primarily for power generation.

Ching River. A total of seventeen dams were planned, primarily to serve irrigation needs on the Ching and its tributaries. The key project is the Ta-fo-szu Dam in Pin Hsien.

Yi/Southern Lo System. Two dams were planned for each main branch of this system to provide for flood control, irrigation, and power to supply industrial growth at Loyang.

Ch'in River. Part of the discharge in the headwaters was to be diverted to the Fen and San-chang rivers. Three major dams were to bring 72 percent of the basin area under flood control and provide irrigation to 2.8 million *mou.*

Notes

Introduction: When the River Runs Clear

1. Gilbert White, *Strategies of American Water Management.*

1. The Problem of Yellow River Management

1. Table 1-1 shows twenty-two rivers with larger drainage basins and at least thirty that carry more water. Several rivers in the United States—the Mississippi, Ohio, St. Lawrence, and Columbia—have larger discharges than the Yellow River, which is comparable to the Tennessee or the Missouri in volume of flow.
2. J. S. Lee, *The Geology of China,* pp. 431–434.
3. S. Ting et al., *Geology and Soils of the Yellow River Basin,* pp. 36–40.
4. Feng Ching-lan, "Special Features of the Physiographic Regions of the Yellow River Basin (Huang-ho Liu-yu Ti-hsing Fen-ch'iu ti T'e-tien)," *KMJP,* Peking, August 1, 1955.
5. V. T. Zaychikov, "Inland Waters," in idem, ed., *The Physical Geography of China (Fizicheskaya Kitaya),* pp. 193–194.
6. Kai J. Hsia and C. S. Lee, "A Study of the Alluvial Morphology of the Lower Wei Basin (Wei-ho Hsia-yu Chung-chi Hsing-t'ai ti Yen-chiu)," *TLHP,* Vol. 29, No. 3 (September 1963), pp. 207–218.
7. O. J. Todd, "Discussion of Saratsi Irrigation Project," *Journal of the Association of Chinese and American Engineers,* May–June 1935. This article is also contained in O. J. Todd, *Two Decades in China.*
8. Zaychikov, "Inland Waters," p. 195.

9. Todd, *Two Decades.*
10. NCNA, Chengchow, July 18, 1958, "River Fighters Guard Yellow River Dikes," *SCMP*, No. 1818 (July 24, 1958), pp. 36–37.
11. Ting et al., *Geology and Soils.*
12. C. H. Kuo, "The Silt of the Yellow River and Its Erosive Action (Huang-ho ti Ni-sha Chi Ch'i Ch'in-shih Tzo-yung)," *TLCS* (August 1956).
13. Discussion between Yellow River Management officials and a U. S. Water Management Delegation, of which the author was a member, at Hua-yuan-k'ou, September 5, 1974.
14. Ts'en Chung-mien, *History of the Changes of the Yellow River (Huang-ho Pien-ch'ien Shih).*

2. Historical Management Strategies

1. See, for example, Karl A. Wittfogel, *Oriental Despotism*; Maurice Meisner, "The Despotism of Concepts: Wittfogel and Marx on China," *China Quarterly*, No. 16 (October–December, 1963), pp. 99–111; and Wu Ta-k'un, "An Interpretation of Chinese Economic History," *Past and Present*, Vol. 1, No. 1 (February 1952), pp. 1–12.
2. Carl Whiting Bishop, "The Beginnings of North and South in China," *Pacific Affairs*, Vol. 7, No. 3 (September 1934), pp. 297–325; Ho P'ing-ti, "The Loess and the Origin of Chinese Agriculture," *American Historical Review*, Vol. 75, No. 1 (October 1969), pp. 1–36; Chi Ch'ao-ting, *Key Economic Areas in Chinese History*; Chang Kuang-chih, *The Archaeology of Ancient China.*
3. Yang K'uan, "Achievements of Hydraulic Engineering in the Warring States Period (Chan-kuo Shih-tai Shui-li Kung-ch'eng ti Ch'eng-chiu)," in Li Kuang-pi and Ch'ien Chun-yeh, eds., *Essays on Chinese Scientific Discoveries and Scientists (Chung-kuo K'e-hsueh Chi-shu Jen-wu Lun-chi).*
4. Ho P'ing-ti, "Loess and Chinese Agriculture," p. 14.
5. Many references to the Cheng-kuo Canal project give the irrigated area as 600,000 acres because of a problem in the conversion from Chinese units of area to English units. The area is always given in Chinese as 40,000 *ch'ing*, and this is translated as 600,000 acres since a modern *ch'ing* is about

fifteen acres. Yang K'uan points out, however, that a *ch'ing* was the equivalent of about five acres when the Cheng-kuo system was built; thus the area was about 200,000 acres.

6. Chi Ch'ao-ting, *Key Economic Areas*, p. 36.
7. Ibid., p. 86.
8. Joseph Needham, *Civil Engineering and Nautics*, Vol. 4, No. 3, of *Science and Civilization in China*, pp. 234–235.
9. Chi Ch'ao-ting, *Key Economic Areas*, p. 10.
10. See, for example, "Remnants of the San-men Gorge Transport Canal (San-men-hsia Ts'au-yun I-chi)," in *Collected Chinese Archaeological Field Reports (Chung-kuo T'ien K'ao-ku Pao-kao Chi)*.
11. Joseph Needham, Vol 4, No. 2 of *Mechanical Engineering*, *Science and Civilization in China*. For recent photographs of wheelbarrows, see *China Pictorial*, April, 1971.
12. Chi Ch'ao-ting, *Key Economic Areas*, pp. 82, 88; Needham, *Civil Engineering*, p. 308.
13. Wittfogel argues that a revolution in social organization (not, fundamentally, in technology), occurred with the management of large-scale irrigation and flood prevention projects. The institutional framework required to mobilize, organize, and direct masses of laborers is said to have evolved into a state mechanism of rigid centralized control, giving rise to a characteristically "Oriental" form of "despotism." Counter arguments hold that the emergence of private land ownership is the significant feature of late Chou times, from which other features, including water management organization, can be explained. See Chapter 2, note 1.
14. Hu Ch'ang-tu, "The Yellow River Administration in the Ch'ing Dynasty," *Far East Review*, Vol. 14 (1955), pp. 505–513.
15. Needham, *Civil Engineering*, pp. 262–263.
16. Quoted in Yang K'uan, "Achievements of Hydraulic Engineering," p. 100.
17. Needham, *Civil Engineering*, p. 267.
18. This description of Chia Jang's strategies, as well as those that follow on other famous managers, is based on Li Yi-chih, "Discussion of the Basic Management Method for the Yellow River (Huang-ho Chih Ken-pen Chih-fa Shang-choi)," *LYCCC*, pp. 401–423, with supplementary information from Chang Han-ying, *Sketch of Historical Yellow River Manage-*

ment Methods (Li-tai Chih-ho Fang-lueh Shu-yao). A brief discussion is also found in Chang Shichin, *General Description of the Yellow River Basin.*
19. Needham, *Civil Engineering,* pp. 247–254.
20. Ch'in Fu, "Famous Treatises," Chapter 7 in *Plans for Regulating Rivers (Chih-ho Fang-lueh).*
21. Hu Ch'ang-tu, "The Yellow River Administration," p. 510.
22. Ibid., p. 506.
23. Ibid., p. 509.

3. Early Western Interest in the Yellow River Problem

1. S. M. Meng, *The Tsungli Yamen: Its Organization and Functions;* Adrian Arthur Bennett, *John Fryer: The Introduction of Western Science and Technology into Nineteenth Century China.*
2. A bibliography of these early writings, prepared by John R. Freeman and held in the Lane Collection at the Water Resources Research Center Library of Colorado State University, is reproduced in Appendix 1 with Freeman's notations on the articles.
3. John R. Freeman, "Flood Problems in China," *Transactions of the American Society of Civil Engineers,* Vol. 75 (1922), pp. 1405–1460.
4. Shen Yi, ed., *Collected Essays on the Yellow River Problem (Huang-ho Wen-t'i T'ao-lun Chi).*
5. See, for example, Freeman's notes on the article by Oxenham in Appendix 1.
6. O. J. Todd, "Repairing the Yellow River Break, Kung Chia Ko, Shantung," *Far East Review,* Vol. 19 (July 1923), pp. 468–481; idem, "The Yellow River Breaks of 1935," *Far East Review,* Vol. 32 (October 1936), pp. 454–461; idem, "The Yellow River Reharnessed," *Geographical Review,* Vol. 39, No. 1 (January 1949), pp. 38–56; O. J. Todd and Siguard Eliassen, "The Yellow River Problem," *Transactions of the American Society of Civil Engineers,* Vol. 105 (1940), pp. 346–453; Todd, *Two Decades.*
7. A. T. Goode et al., *Report of the Committee of Experts on Hydraulic and Road Questions in China;* Robert Haas, *Report of the Secretary of the Council Committee on His Mission in*

China; L. Rajchman, *Report of the Technical Agent of the Council on His Mission in China*.

8. W. C. Lowdermilk, "Man-Made Deserts," *Pacific Affairs*, Vol. 8, No. 3 (December 1935), pp. 409–419; idem, "Forestry in Denuded China," *Annals of the American Academy of Political and Social Science*, Vol. 152 (November 1930), pp. 127–141; James Thorp, *Geography of the Soils of China*.

9. Eugene Reybold, James P. Growden, and John L. Savage, *Preliminary Report on the Yellow River Project*.

10. John S. Cotton, *Preliminary Report on the Yellow River Project*.

11. For a synopsis of the Japanese plan see Chang Shichin, *General Description of the Yellow River Basin*. For an evaluation of this plan see Appendix 3.

12. *A List of Universities, Independent Colleges, and Technical Schools in China, May, 1935*.

13. For a full biography of Li see Sung Hsi-shang, *Biography of Li Yi-chih (Li Yi-chih Ti Sheng-p'ing)*.

14. Shen Yi, *Collected Essays*.

15. See, for example, the following articles in *LYCCC*: "Outline for Yellow River Management Work (Chih-li Huang-ho Kung-tso Kang-yao)," pp. 458–463; "Opinion on Yellow River Management (Chih-huang Yi-chien)," pp. 463–465; "An Opinion on Yellow River Management (Kuan-yu Huang-ho Chih-tao Chih Yi-chien)," pp. 465–467.

16. Li Yi-chih, "Work Plan for the Yellow River Water Conservancy Commission (Huang-ho Shui-li Wei-yuan-hui Kung-tso Chi-hua)," *LYCCC*, pp. 450–458.

17. Li Yi-chih, "Discussion of the Basic Management Method for the Yellow River," *LYCCC*, pp. 401–402.

18. *LYCCC*, pp. 204–213 and 213–216 respectively.

19. The China International Famine Relief Commission surveys, for example, included work in the Great Bend and the Fen Valley, as well as a flood control survey in Shantung. See Todd, *Two Decades*, and Andrew James Nathan, *A History of the China International Famine Relief Commission*.

20. The information for this section on soil and water conservation was accumulated from W. H. Huang, *Soil and Water Conservation of the Yellow River Basin*, and various plans and reports from the experiment stations themselves held at the Water Resources Planning Commission, Taipei. See Bibliography.

21. Li Yi-chih, "General Overview of a Yellow River Management Plan (Huang-ho Chih-pen Chi-hua Kai-yao Su-mu)," *LYCCC*, pp. 441–450. For a summary see Chang Shichin, *General Description of the Yellow River Basin*.
22. See Reybold, Growden, and Savage, *Preliminary Report*.
23. Cotton, *Preliminary Report*.
24. Chang Shichin, *General Description of the Yellow River Basin*.
25. Ibid.
26. These are the figures shown in Table 3-3, but Chang Shichin, *General Description of the Yellow River Basin*, gives about 1.5 million *mou* under irrigation in 1947, instead of 2.6 million *mou*. Such discrepancies only illustrate the unreliability of information in the hands of different engineers at the time, which in itself indicates how much groundwork was required before project construction could be undertaken.
27. WRPC No. 59.
28. Ministry of Water Conservancy (Shui-li Pu), *Water Conservancy Administration (Shui-li Hsing-cheng)* (SLHC).
29. WRPC Nos. 100, 106, 121.

4. Soviet Assistance in Yellow River Development

1. Teng Tse-hui, *Report on the Multiple Purpose Plan for Permanently Controlling the Yellow River and Exploiting Its Water Resources*.
2. Chi Wen-shun, "Water Conservancy in Communist China," *China Quarterly*, No. 23 (July–September 1965), pp. 37–54.
3. See, for example, Ch'en Chao-ping, "What We Have Learned from the Soviet Experts (Wo-men Ts'ung Su-lien Chuan-chia-men Hsueh-hsi So Shem-me)," *CKSL*, No. 11 (November 1957), pp. 8–9; Liu Te-jen, "The Assistance of Soviet Experts in Project Management Work (Su-lien Chuan-chia Tui Kung-ch'eng Kuan-li Kung-tso ti Pang-chu)," Ibid., pp. 6–7; T'ang Shih, "What the Soviet Experts Have Contributed to China's Power Industry in the Past Eight Years," *People's Power Industry (Jen-min Tien-yeh)*, No. 31 (November 5, 1957); *ECMM*, No. 116 (January 27, 1958), pp. 34–38.
4. The results of expeditions to the headwaters in 1952, for example, showed that Chinese maps previously had confused the

locations of the twin lakes, Oring Nor and Jaring Nor, and had not depicted the true source area of the river correctly. NCNA, Kaifeng, January 7, 1953, "Exact Source of Yellow River Thoroughly Explored," *SCMP*, No. 487 (January 8, 1953), pp. 20–21.

5. NCNA, Chengchou, October 25, 1956, "Hydrological Survey along Yellow River," *SCMP*, No. 1400 (October 30, 1956), p. 16.

6. NCNA, Chengchou, April 25, 1955, "Exhibition on Yellow River Control Opened in Chengchou," *SCMP*, No. 1036 (April 28, 1955), p. 52.

7. Lan K'o-chen, "On the Northward Transfer of Southern Waters (Lun Nan-shui Pei-tiao)," *TLCS* (April 1959), pp. 145–146.

8. A. A. Koroliev, "Problems Concerning Future Development of China's Water Conservancy Work (Kuan-yu Chung-kuo Shui-li Shih-yeh Chin-hou Fa-chan ti Mou-hsieh Wen-t'i)," *SLYTL*, No. 4 (April 1959), pp. 19–32; *ECMM*, No. 168 (May 18, 1959), pp. 25–36.

9. The best single source on the entire question of interbasin transfer is Wang Hua-yun, "The Great Rationality of Transferring Southern Waters Northward (Nan-shui Pei-tiao ti Hsiung-wei Li-hsiang)," *Red Flag (Hung Ch'i)*, September 1, 1959, pp. 36–44. See also T'ao Tuan-shih, "The Routes Question in Diversion of the Han, Yellow, and Huai Rivers (Yin-han Chi-huang Chi-huai Lu-hsien Wen-t'i)," *JMJP*, June 4, 1956; *Ajia no Yume*; and Lan K'o-chen, "On The Northward Transfer."

10. The details on exactly how and where this canal would enter the Yellow River system were not made available. The problem most likely had not been solved when these preliminary route selections were made. Wang Hua-yun, "The Great Rationality," simply indicates Ting-hsi as the end point of the canal. Other sources indicate that three tributaries, the T'ao River, the upper Wei River, and the Tsu-li River, all were possibilities for feeding water into the Yellow River. See *HKWHP*, May 20, 1959, "The Northward Transfer of Southern Waters Has Two Dragons, Boring through Mountains and Going over Peaks to Bring Water (Nan-shui Pei-tiao Yu Shuang-lung Ch'uan-shan Yueh-ling Tsai Shui Lai);" NCNA, Chengchou, October 18, 1958, "Preliminary Survey of Yangtze–Yellow River Canal Completed," *SCMP*, No.

1880 (October 23, 1958), p. 23; NCNA, Chengchow, October 21, 1958, "Another Yangtze–Yellow River Canal Planned," *SCMP*, No. 1881 (October 24, 1958), pp. 25–26.

11. Wang Hua-yun, "The Great Rationality."

12. Discussion between Yellow River officials and a U.S. Water Management Delegation, of which the author was a member, in Chengchou, September 4, 1974.

13. Koroliev, "Problems Concerning Future Development."

14. Minor transfers, both in and out of the Yellow River basin, apparently are made. Discussion with officials at the Chiang-tu Pumping Station near Nanking on September 2, 1974, indicated that some water from the Huai River basin occasionally goes northward via the Grand Canal, and there has been reference to 270 million cubic meters diverted northward from near Chengchou to relieve drought in the Tientsin vicinity in May 1973. See *JMJP*, December 20, 1974a, "Reconstruction of Key San-men Gorge Water Conservancy Project Attains Initial Success (San-men-hsia Shui-li Shu-niu Kung-ch'eng Kai-chien Huo-te Ch'u-pu Ch'eng-kung)."

15. The following discussion is based on *Ajia no Yume*, pp. 73–97.

16. The dam at Wang-wang-chuang, farthest downstream of the major dams, was not among the original forty-six, but plans for its construction were later developed, bringing the total to forty-seven. See NCNA, Tsinan, January 23, 1960, "Another Dam Being Built on Lower Reaches of Yellow River," *SCMP*, No. 2186 (February 1, 1960), p. 10; and *CNS*, Tsinan, January 10, 1960, "Construction Begins on a Large-Scale Water Conservancy Project in Shantung (Shan-tung Yi-ko Ta-hsing Shui-li Kung-ch'eng Tung-kung Hsing-chien)."

17. See Teng Tse-hui, *Report on the Multiple Purpose Plan*. The San-sheng-kung project is referred to as being located near Teng-k'ou, Inner Mongolia, in Teng's report.

18. See NCNA, Lanchou, September 28, 1958, "Two New Multipurpose Projects along the Yellow River," *SCMP*, No. 1867 (October 3, 1958), p. 32; NCNA, Tsinan, December 15, 1959, "Yellow River Dam in Lower Reaches Completed," SCMP, No. 2161 (December 22, 1959), p. 31; NCNA, Tsinan, January 3, 1960, "Yellow River Dam with Sorghum Stalks and Earth," *SCMP*, No. 2172 (January 8, 1960), p. 15; NCNA, Tsinan, February 27, 1960, "Construction of New Yellow

River Harnessing Project," *SCMP*, No. 2208 (March 3, 1960), p. 12; *CNS*, January 10, 1960, "Construction." *Ajia no Yume* reports that construction was begun on thirteen additional dams through 1960, but without confirmation from Chinese sources it can only be assumed that this may refer to surveying or other work preliminary to actual construction. The thirteen are: the Lung-yang, Szu-k'ou, Pa-pan and Ts'ai-chia gorges above Lanchou; Wu-chin Gorge, Hei-shan Gorge and Ta-liu-shu between Lanchou and Ch'ing-t'ung Gorge; Wan-chia-sai, Yu-k'ou, Ch'ien-pei-hui, Lo-ku-k'ou, and Chi-k'ou along the upper portion of the Shansi-Shensi border stretch of the river; Pa-li-hu-t'ung below San-men Gorge. *Ajia no Yume* also mentions another dam for which plans were announced subsequent to the original report. Located above Lung-yang Gorge near Chung-ho, Chinghai Province the project is called Wai-ssu, but the lack of reference to this project in any Chinese publications makes its existence questionable.

19. *JMJP*, September 16, 1974a, "Key Ch'ing-t'ung Gorge Water Conservancy Project Basically Completed (Ch'ing-t'ung-hsia Shui-li Shu-niu Kung-ch'eng Chi-pen Chien-ch'eng)."

20. Kansu Provincial Radio, September 24, 1974, "Kansu Completes Yellow River Hydroelectric Plant," FBIS, September 26, 1974.

21. *JMJP*, February 5, 1975, "Our Country's Biggest Hydroelectric Power Station—Liu-chia Gorge Hydroelectric Power Station—Victoriously Completed (Wo-kuo Tsui-ta-ti Shui-tien-chan—Liu-chia-hsia Shui-tien-chan—Sheng-li Chien-ch'eng)."

22. *JMJP*, December 20, 1974a; *JMJP*, December 20, 1974b, "Understand the Yellow River, Rebuild the Yellow River (Jen-shih Huang-ho, Kai-chien Huang-ho)."

23. The best summary of positive and negative features of early construction of the dam at San-men Gorge is provided in Wang Hua-yun, "The Great Significance of the Key Project at San-men Gorge (Huang-ho San-men-hsia Shui-li Shu-niu Kung-ch'eng ti Chung-yao Yi-yi)," *CKSL*, No. 15 (March 1957), pp. 1–4. The discussion that follows is based largely on the points Wang makes. Numerous other articles on various planning and technical aspects of the decision to build

at San-men Gorge were published, and many are collected
in the entire issues of *CKSL*, Nos. 15 (March 1957), 19 (July
1957), and 20 (August 1957).

24. The site was chosen over an alternative plan that would have
put two large dams in the same part of the basin—one at
Chih-ch'uan upstream near the mouth of the Wei River, and
one at Mang-shan downstream near Chengchou. See Chun
Ch'ien, "The Outstanding Contribution of the Soviet Advisors in the San-men Gorge Project (Su-lien Chuan-chia
Tui San-men-hsia Shui-li Shu-niu Kung-ch'eng ti Cho-yueh
Kung-hsien)," *JMJP*, April 15, 1957; Liu Tze-hou, "The Assistance of Soviet Experts in the San-men Gorge Project
(Su-lien Chuan-chia Tui San-men-hsia Shui-li Kung-ch'eng
ti Pang-chu)," *Wen-hui Pao*, Shanghai, November 6, 1957.

25. Wang Hua-yun, "Great Significance."

26. Plans to make the dam even higher and raise the normal high
water level to 370 meters also were considered. See Shen
Ts'ung-kang, "On the Planning Situation for the First Stage
at San-men Gorge (San-men-hsia Ch'u-pu She-chi Ch'ing-
kuang Chien-shao)," *CKSL*, No. 19 (July 1957), pp. 11–15.

27. NCNA, San-men Gorge, March 29, 1957, "Construction of San-
men Gorge Will Start Soon," *SCMP*, No. 1503 (April 3, 1957),
pp. 13–14; NCNA, San-men Gorge, April 15, 1957, "Machines Get into Action at San-men," *SCMP*, No. 1514 (April
23, 1957), p. 4.

28. NCNA, Lanchou, September 28, 1958.

29. NCNA, Peking, March 13, 1958, "San-men Gorge Project to be
Completed Ahead of Schedule," *SCMP*, No. 1735 (March 20,
1958), pp. 7–8; NCNA, San-men Gorge, June 6, 1958, "San-
men Gorge Project Going Ahead at Top Speed," *SCMP*, No.
1791 (June 13, 1958), p. 13; NCNA, Chengchou, August 8,
1959, "New Stage in Yellow River Multiple Purpose Project,"
SCMP, No. 2075 (August 13, 1959), pp. 25–26; NCNA,
Chengchou, September 16, 1959, "Work Begins on Big Yellow River Hydro-electric Power Station," *SCMP*, No. 2100
(September 22, 1959), p. 29; NCNA, Chengchou, October 3,
1959, "China's Sorrow Is Being Tamed," *SCMP*, No. 2131
(November 5, 1959), pp. 28–30; *SJP*, Sian, November 26,
1959, "Resettlement of People in the Area of the San-men
Gorge Reservoir, Shensi, Commences," *SCMP*, No. 2173
(June 11, 1960), pp. 16–18; NCNA, San-men Gorge, Decem-

ber 10, 1959, "Builders of San-men Gorge Dam on Yellow River Complete 1959 Target," *SCMP*, No. 2157 (December 21, 1959), p. 36.

30. Lei Chien, "Desolate San-men Gorge (Leng-lo-le ti San-men-hsia)," *HTJP*, July 27 and July 28, 1963.

31. NCNA, Chengchou, March 19, 1961, "Big Irrigation System Expanded," *SCMP*, No. 2463 (March 24, 1961), p. 17; NCNA, Sian, January 3, 1962," Extensive Antierosion Work in Yellow River Basin," *SCMP*, No. 2654 (January 6, 1962), pp. 25–26; *KJJP*, Peking, January 5, 1962, "New Face of San-men Gorge (San-men-hsia Hsin-mao)." There has been no denial or confirmation of reports in newspapers antagonistic to the Chinese government that part of the dam was blasted to accommodate a flood crest which the normal spillways could not accommodate. See Meng Lin, "News of the Communists Blasting San-men Gorge Reservoir (Chung-kung Tza-hui San-men-hsia Shui-k'u Chi-wen)," *CCNW*, December 16, 1964, and Meng Lin, "More on the Matter of the Communists Blasting the San-men Gorge Reservoir (Tsai T'an Chung-kung Tza-hui San-men-hsia Shui-k'u Shih-chien)," *CCNW*, March 16, 1965.

32. *JMJP*, December 20, 1974a; *JMJP*, December 20, 1974b.

33. Discussions between Yellow River officials and an American Water Management Delegation, of which the author was a member, Chengchou, September 4, 1974.

34. *HTJP*, November 11, 1951, "Construction Completed on Huang-Yang Gate Irrigation Project in the Ho-t'ao Region of Sui-yuan (Sui-yuan Ho-t'ao Huang-yang-cha Kuang-kai Kung-ch'eng Yi Ts'ao-ch'eng)"; Ch'in Hsu-lun, "The Great Bend of the Yellow River Today (Chin-jih ti Huang-ho Ho-t'ao)," *HKTKP*, June 7, 1955; NCNA, Shan-pa, Inner Mongolia, June 30, 1957, "Projects to Reconstruct Irrigation Channels along Pao-t'ou–Lanchou Railway in Inner Mongolia Completed," *SCMP*, No. 1569, July 15, 1957, p. 25; Shen Su-ju, "At the Great Bend of the Yellow River," *China Reconstructs* (July 1964), pp. 32–35.

35. *JMJP*, September 16, 1974b, "Long Journey on the Yellow River (Huang-ho Wan-li Hsing)."

36. Discussion with Chinese water management officials, Peking, September 12, 1974.

37. See note 16.

38. NCNA, Hong Kong, October 31, 1966, "Teng-k'ou Electric Pumping Station in Inner Mongolia is Completed and Begins Pumping (Nei-meng-ku Teng-k'ou Tien-li Yang-shui-chan Wan-kung Ch'ou-shui)."

39. *JMJP*, April 9, 1958, "Construction Begun on Kang-li Yellow River Diversion Irrigation Project (Kang-li Yin-huang Kuang-kai Kung-ch'eng K'ai-kung)."

40. NCNA, Chengchou, December 2, 1959, "Work Begins on Seventh Giant Yellow River Reservoir," *SCMP*, No. 2151 (December 8, 1959), p. 40.

41. These visits were made on September 4 and September 5, 1974, by the U. S. Water Management Delegation.

42. NCNA, Chengchou, June 3, 1957, "Progress of Water and Soil Conservation Work along Upper and Middle Reaches of Yellow River Criticized," *SCMP*, No. 1552 (June 18, 1957), pp. 23–24.

43. *JMJP*, March 23, 1962, "Reclamation of Wasteland on Steep Slopes Must Be Restricted," *SCMP*, No. 2719 (April 16, 1962), pp. 15–16.

44. M. F. Chao, 1963a, "Water and Soil Conservation in Regions along the Middle Reaches of the Yellow River," *SLYTL*, No. 13 (July 5, 1963), *SCMM*, No. 380 (September 3, 1963), pp. 1–8.

45. Fang Cheng-san, "An Exploration into the Question of Water and Soil Conservation on the Loess Highland in the Northwest," *JMJP*, May 24, 1965, *SCMP*, No. 3480 (June 18, 1965), pp. 7–11; Li Fu-tu, "An Important Way to Solve the Problem of Silt and Sand in the Yellow River," *JMJP*, January 26, 1966.

46. For a discussion of the powers and responsibilities of the State Council Staff Offices, see A. Doak Barnett, *Cadres, Bureaucracy, and Political Power in Communist China*.

47. *JMJP*, September 9, 1965, "National Conference on Water Conservation Stresses that Water Conservation Should Render Better Service for Increasing Agricultural Production," *SCMP*, No. 3545, April 8, 1965, pp. 11–13.

48. The advantages of using local decision-making agencies in cases where the river basin framework is not appropriate are discussed by White, *Strategies of American Water Management*, and Leonard R. Brown, "Are River Basins the Best

Framework for Water Resources Planning, Development, and Management?," *Water Resources Bulletin*, Vol. 8, No. 2 (April 1972), pp. 401–403.

5. Reliance on Traditional Methods

1. Yeh Yung-yi, "Some Special Points on Solving the Flood Problem in the Yellow River Management Plan (Chih-li Huang-ho Kuei-hua Chung Chieh-chueh Fang-hung Wen-t'i Chi-ke T'e-tien)," *KMJP*, August 4, 1955; *JMJP*, March 31, 1955, "Huge, Temporary Flood Prevention Project Built on Yellow River Lower Reaches (Huang-ho Hsia-yu Hsing-hsiu Chu-ta ti Lin-shih Fang-hung Kung-ch'eng)."
2. *HCJP*, March 22, 1958, "Ten Thousand Workers Take Part in River Dredging Project (Wan-jen Ts'an-chia Wa-ho Kung-ch'eng)."
3. NCNA, Peking, April 4, 1954, "Dikes along Yellow River Being Strengthened," *SCMP*, No. 783 (April 7, 1954), p. 8; NCNA, Peking, February 9, 1962, "Water Conservancy Projects Built along Yellow and Yangtze Rivers," *SCMP*, No. 2679 (February 16, 1962), pp. 20–21; *JMJP*, November 11, 1963, "Winter Reconstruction Projects Successively Begin Work on Lower Yellow River (Huang-ho Hsia-yu Tung-hsiu Kung-ch'eng Lu-hsu K'ai-kung)."
4. NCNA, Chengchou, July 18, 1958, "River Fighters Guard Yellow River Dikes," *SCMP*, No. 1818 (July 24, 1958), pp. 36–37; NCNA, July 23, 1958, "Yellow River Crest Safely Passes Hazardous Section," *SCMP*, No. 1821 (July 29, 1958), p. 15.
5. NCNA, Peking, June 28, 1956, "Yellow River Water Crest Approaches Coast," *SCMP*, No. 1324 (July 9, 1956), p. 5.
6. NCNA, Peking, June 24, 1971, "Achievements in Using Yellow River Water for Irrigation," *SCMP*, No. 4929 (July 2, 1971), pp. 211–214.
7. NCNA, Peking, May 28, 1954, "Fu Tso-yi Reviews Water Conservancy Work in China: Irrigation Aids Agriculture," *SCMP*, No. 821 (June 3, 1954), pp. 29–30.
8. NCNA, Peking, August 11, 1955, "Yellow River Control Work Reviewed," *SCMP*, No. 1108 (August 12, 1955), pp. 15–16.
9. Han Fang, "Yellow River Water and Hua-yuan-k'ou (Huang-ho

Shiu Yu Hua-yuan-k'ou)," YCWP, Kuang-chou, August 25, 1965.

10. NCNA, Peking, December 21, 1957, "Yellow River Tributary Basin Curtailed," SCMP, No. 1680 (December 30, 1957), pp. 7–8.

11. NMKJP, July 28, 1956, "A Strong Yellow River Levee (Yi-tao Chien-ku ti Huang-ho T'i)."

12. CKHW, Kuang-chou, April 13, 1962, "Yellow River Irrigation District Increases Spring Repairs (Huang-ho Kuang-ch'u Cha-fan Ch'un-hsiu)."

13. NCNA, Tsinan, July 22, 1958, "Yellow River Water Diverted by Lake," SCMP, No. 1820 (June 28, 1958), p. 19; NCNA, Chengchou, August 5, 1958, "Meeting Plans to Strengthen Yellow River Projects," SCMP, No. 1831 (August 13, 1958), pp. 37–38.

14. Chun Ch'ien, "Work Completed on Shih-t'ou-chuang Overflow Weir (Shih-t'ou-chuang Yi-hung-yen Wan-kung)," JMJP, September 10, 1951; Shao Tang, "Overflow Weir Project to Control the Yellow River (Chih-li Huang-ho ti Yi-hung-yen Kung-ch'eng)," HKTKP, July 11, 1954.

15. NCNA, Chengchou, May 19, 1956, "Yellow River Diversion Scheme in Honan," SCMP, No. 1295 (May 24, 1956), p. 25; NCNA, Chengchou, June 29, 1956, "Yellow River Water Diversion Project Completed," SCMP, No. 1324 (July 9, 1956), pp. 14–15.

16. NCNA, Kaifeng, July 4, 1953, "Yellow River Headquarters Set Up," SCMP, No. 604 (July 7, 1953), p. 7; NCNA, Chengchou, May 28, 1958, "Conference of Yellow River Antiflood Headquarters Decides to Strive for Greatest Victory Possible," SCMP, No. 1790 (June 12, 1958), pp. 18–19.

17. NCNA, Chengchou, July 1, 1956, "Yellow River Flood Prevention Headquarters Issues Directive on Possible Arrival of Larger Floods," SCMP, No. 1329 (July 10, 1956), pp. 18–19; NCNA, Peking, July 20, 1958, "Yellow River Torrent Reaches Coastal Shantung Province," SCMP, No. 1819 (July 25, 1958), pp. 9–10.

18. NCNA, Kaifeng, April 6, 1953, "Wheat Fields Irrigated by Yellow River in Honan," SCMP, No. 545 (April 5–7, 1953), p. 31.

19. Meng Chin-chih, "The New Face of the Yellow River (Huang-ho ti Hsin Mien-mao)," HKWHP, August 23, 1966.

20. NCNA, Chengchou, March 15, 1959, "China's Sorrow Now Serves Farmers," *SCMP*, No. 1975 (March 18, 1959), pp. 36–37.

21. *HTJP*, November 11, 1951; Ch'in Hsu-lun, "The Great Bend of the Yellow River Today (Chin-jih ti Huang-ho Ho-t'ao)," HKTKP, June 7, 1955; Shen Su-ju, "At the Great Bend."

22. NCNA, Shan-pa, Inner Mongolia, June 30, 1957.

23. NCNA, Lanchou, September 19, 1955, "New Irrigation Canal along Yellow River Reaches," *SCMP*, No. 1133 (September 20, 1955), p. 53; NCNA, Yin-ch'uan, December 31, 1959, "New Yellow River Irrigation Canal in Northwest China," *SCMP*, No. 2171 (January 7, 1960), p. 27; *JMJP*, May 12, 1956, "Spring Repair Project Completed on Yin-ch'uan Special District Canal (Yin-ch'uan Ch'uan-ch'u Wan-ch'eng Ch'u-tao Ch'un-hsiu Kung-ch'eng)"; Shen Su-ju, "At the Great Bend."

24. Shen Su-ju, 1964, "At the Great Bend," p. 34.

25. NCNA, Hong Kong, June 30, 1964, "Ninghsia Yellow River Irrigation District Builds a Permanent Flood Prevention Levee (Ning-hsia Yin-huang Kuang-ch'u Hsiu-chien Yung-chiu-hsing Fan-hung Kung-ch'eng)."

26. Teng Tse-hui, *Report on the Multiple Purpose Plan*, p. 39.

27. A description of these failures is given in *KMJP*, January 11, 1957, "Soil Conservation Suffers on Middle and Downstream of Yellow River," *SCMP*, No. 1455 (January 22, 1957), pp. 13–14. The antierosion progress reports from the 1950s are very difficult to use since figures sometimes were given for a certain area as under "basic erosion control," and sometimes under "partial control," or "initial control." Often reports of planned projects were more specific than reports of completed projects. And in the early years these almost invariably were overoptimistic. Shensi Province, for example, announced plans in 1956 to end all soil and water loss due to erosion within seven years. NCNA, Sian, July 22, 1956, "Water and Soil Conservation Work in Shensi," *SCMP*, No. 1336 (July 25, 1956), pp. 14–15. Other descriptions of early antierosion work can be found in Chao Ming-fu, "A Discourse on Soil Conservancy in the Yellow River Basin," *Red Flag*, No. 21 (November 21, 1962), *SCMM*, No. 341 (November 26, 1962), pp. 29–37, and Chao Ming-fu, "Water

and Soil Conservation along the Middle Reaches of the Yellow River," *SLYTL*, No. 13 (July 5, 1963), *SCMM*, No. 380 (September 3, 1963), pp. 1–8.

28. NCNA, Peking, September 22, 1959, "Mass Campaign to Control Soil Erosion to Be Launched in Yellow River Basin," *SCMP*, No. 2104 (September 28, 1959), p. 30.

29. NCNA, Peking, November 17, 1959, "Yellow River Water and Soil Conservation Projects," *SCMP*, No. 2142 (November 24, 1959), p. 7.

30. NCNA, Sian, January 3, 1962, "Extensive Antierosion Work in Yellow River Basin," *SCMP*, No. 2654 (January 9, 1962), pp. 25–26; NCNA, Taiyuan, August 19, 1963, "Antierosion Measures by People's Communes in Yellow River Province," *SCMP*, No. 3045 (August 22, 1963), pp. 11–12; NCNA, Sian, October 13, 1964, "Antierosion Work in Yellow River Basin," *SCMP*, No. 3321 (October 21, 1964), p. 18; NCNA, December 1, 1965, "People Make Semibarren Land in Northwest China Yield Stable Harvests," *SCMP*, No. 3591 (December 6, 1965), p. 21.

31. NCNA, Sian, June 15, 1965, "Massive Antierosion Project in Middle Reaches of Yellow River," *SCMP*, No. 3480 (June 18, 1965), p. 18.

32. Chao Ming-fu, "Provide Stronger Leadership and Rely on the Masses in Bringing About a New Upsurge in the Water and Soil Conservation Work in the Middle Reaches of the Yellow River," *SLYTL*, No. 24 (December 20, 1963), *SCMM*, No. 403 (February 10, 1964), pp. 6–11.

33. NCNA, Peking, October 31, 1971, "Mass Efforts at Soil Conservation Help to Control Yellow River," *SCMP*, No. 5011 (November 10, 1971), pp. 116–118; *KMJP*, February 6, 1971, "Vigorously Make a Success of Water and Soil Conservation, Develop Agricultural Production in Mountain Areas," *SCMP*, No. 484 (February 19, 1971), pp. 134–141.

34. *JMJP*, December 20, 1974a and b.

35. NCNA, Chengchou, October 22, 1965, "China's Scientists Work to Increase Soil Fertility in Yellow River Valley," *SCMP*, No. 3588 (December 1, 1965), p. 19; *JMJP*, May 23, 1965, "Li-ch'eng Cleverly Uses Yellow River Benefits and Cleverly Eliminates Silt Water Disasters (Li-ch'eng Ch'iao-yung Huang-ho Chih-li Ch'iao-pi Huang-shui Chih-hai)"; Huang

Li, "Can the Yellow River Irrigate the Land (Huang-ho Shui K'o-pu-k'o-yi Jao-ti)," *YCWP*, July 12, 1965.

36. NCNA, Peking, June 24, 1971.

37. *JMJP*, May 28, 1971, "Drawing Yellow River Water for Irrigation, Transforming Harm Into Blessing," *SCMP*, No. 4918 (June 15, 1971), pp. 53–63.

38. *JMJP*, September 25, 1957, "Raise an Enthusiastic Wave of Water Conservancy Construction for Agricultural Land," *SCMP*, No. 1626 (October 8, 1957), pp. 16–18.

39. *JMJP*, September 12, 1958, "CCP Central Committee Directive on Water Conservancy Work," *SCMP*, No. 1857 (September 19, 1958), pp. 4–6.

40. The National Agricultural Development Program, approved in April 1960, emphasized water management methods to some agriculture, and the new economic policy outlined at the second National Peoples Congress in 1962 gave agriculture a development priority over both heavy and light industry, and called for cutbacks in capital construction generally.

41. Liu Yeh, "The Lifeline of Production in Mountainous Areas," *JMJP*, January 18, 1962, *SCMP*, No. 2674 (February 8, 1962), pp. 4–15; Tung Keng, "Winter Water Conservation Works Show Marked Results," *Current Events Handbook (Shih-Shih Shou-ts'e)*, No. 2 (January 16, 1965), *SCMM*, No. 460 (March 16, 1965), pp. 24–25; Liu Shen, "Tentative Views on Some Basic Questions in Our Country's Water Conservancy Construction," *Economic Research (Ching-chi Yen-chiu)*, No. 6 (June 20, 1965), *SCMM*, No. 483 (August 9, 1965), pp. 24–32.

42. Meeting with Yellow River management officials, Chengchou, September 4, 1974. Another report in *JMJP*, September 16, 1974b, states that the total of such small diversions is 179.

43. *JMJP*, May 23, 1965; *JMJP*, May 26, 1971, "Utilize Water and Sand Resources of the Yellow River, Build Stable-Yield and High-Yield Farm Fields," *SCMP*, No. 4914 (June 9, 1971), pp. 117–122.

44. Visit to Mang-shan Pumping Station, September 4, 1974.

45. Meeting with Yellow River management officials, Chengchou, September 4, 1974.

46. *JMJP*, May 28, 1971.

47. NCNA, Chengchou, October 3, 1959.
48. NCNA, Taiyuan, January 12, 1959, "Junks Sail on Mountainous Area in Shansi," *SCMP*, No. 1934 (January 15, 1959), p. 22.
49. NCNA, Hong Kong, September 17, 1963, "Ninghsia Yellow River Irrigation District Builds Three New Electrified Drainage Stations (Ning-hsia Huang-ho Kuang-kai-ch'u Hsin-chien San-tso Tien-li Pai-shiu-chan)."
50. NCNA, Hong Kong, May 10, 1964, "Ninghsia Yellow River Irrigation District Builds a New Drainage Project (Ning-hsia Yin-huang Kai-ch'u Hsin-chien Yi-p'i P'ai-shui Kung-ch'eng)."
51. NCNA, Hong Kong, October 11, 1963, "Shensi-Ninghsia Canal Siphon Projects Progress Well (Ch'in-ning Ch'u-tao Hsi-kuan Kung-ch'eng Chih-liang Hao)."
52. *JMJP*, November 11, 1964, "Build a Drainage Network, a Year Round Remedy to Waterlogging (Chien-chii P'ai-shui-kang Ch'uan-nien Wu Nei-loa)."
53. Hsieh Shih-yen and Chang Chih-ch'u, "A Visit to Ch'ing-t'ung Gorge (Yao-fang Ch'ing-t'ung-hsia)," *JMJP*, January 11, 1960; *NCNA*, Yin-ch'uan, January 11, 1959, "Irrigation Network Being Built in Southern Ninghsia," *SCMP*, No. 1934 (January 15, 1959), p. 24.
54. NCNA, Yin-ch'uan, August 22, 1959, "First Stage of Big Reservoir Finished on Yellow River Tributary," *SCMP*, No. 2085 (August 27, 1959), p. 30.
55. NCNA, Sian, November 17, 1962, "Electric Pumps Lift Water Seventy Meters in Northwest China," *SCMP*, No. 2864 (November 21, 1962), p. 10.
56. "Water for the Loess Highlands," *China Reconstructs* (August 1970), p. 22.
57. *TKP*, Peking, September 14, 1958, "Brave People and the Bravery Canal (Ying-hsuing Jen-min Ying-hsing Ch'u)."
58. NCNA, Peking, December 1, 1965.
59. NCNA, Taiyuan, April 16, 1970, "Commune in Shansi Relies on Itself in Water Conservancy Construction," *SCMP*, No. 4643 (April 27, 1970), pp. 21–23.
60. NCNA, Sian, September 10, 1965, "Northwest China Exploits Underground Water for Irrigation," *SCMP*, No. 3537 (September 15, 1965), p. 17.
61. NCNA, Huhehot, May 5, 1956, "Nine Thousand Volunteer Land

Reclaimers Arrive in Inner Mongolia," *SCMP*, No. 1302 (June 5, 1956), p. 24.

62. Shen Su-ju, "At the Great Bend"; NCNA, Hong Kong, September 9, 1963, "New Face of the Seventy-Two Connected Lakes (Ch'i-shih-erh Lien-hu Hsin-mao)."

63. NCNA, Lanchou, November 2, 1954, "Yellow River Soil Erosion Can Be Checked, Scientists Say," SCMP, No. 921 (November 3, 1954), pp. 11–12.

64. Combined Investigation Team on Soil and Water Conservation in the Middle Yellow River, Chinese Academy of Sciences, eds., *Soil and Water Conservation Handbook (Shui-t'u Pao-ch'ih Shou-ts'e)*.

65. NCNA, Sian, January 3, 1962; NCNA, Taiyuan, August 19, 1963; NCNA, Sian, October 13, 1964; NCNA, Sian, June 15, 1965; Chao Ming-fu, "Water and Soil Conservation."

66. Chao Ming-fu, "Water and Soil Conservation."

67. NCNA, Yin-ch'uan, June 5, 1959, "Experiments to Improve Alkaline Soils in Northwest China," *SCMP*, No. 2032 (June 11, 1959), pp. 24–25; NCNA, Huhehot, October 10, 1961, "Deserts in Inner Mongolia Being Tamed," *SCMP*, No. 2599 (October 17, 1961), pp. 19–20; NCNA, Yin-ch'uan, April 24, 1966, "Chinese Desert Control Workers Find New Way of Improving Extending Plant Cover," *SCMP*, No. 3686 (April 28, 1966), p. 13.

68. NCNA, Yin-ch'uan, June 5, 1959.

69. *JMJP*, September 16, 1974b.

70. Meeting with Chinese water management officials, Peking, September 12, 1974.

71. Ibid.

6. The New Strategy and Its Implications

1. Owen Lattimore, *Inner Asian Frontiers of China*, p. 512.

2. Henry G. Schwarz, "Chinese Migration to Northwest China and Inner Mongolia, 1949–1959," *China Quarterly*, No. 16 (October–December 1963), pp. 62–74.

3. Chang Kuei-sheng, "Nuclei Formation of Communist China's Iron and Steel Industry," *Annals of the Association of*

American Geographers, Vol. 60, No. 2 (June 1970), pp. 257–
285.
4. Wu Yuan-li, *The Spatial Economy of Communist China.*
5. Ch'in Fu, *Plans for Regulating Rivers (Chih-ho Fang-lueh).*
6. June Mary Zaccone, "Some Aspects of Surplus Labor, Water
Control, and Planning in China, 1949–1960," Ph.D. dissertation, University of North Carolina, 1963.
7. Lynn White, Jr., "The Historical Roots of Our Ecological Crisis,"
Science, Vol. 155 (1967); Lewis W. Moncrief, "The Cultural
Basis for Our Environmental Crisis," *Science*, Vol. 170, No.
3957 (October 30, 1970), pp. 508–512.
8. Leo Marx, *The Machine in the Garden*; Lewis Mumford, *Technics and Civilization.*
9. M. Taghi Farvar and John P. Milton, eds., *The Careless Technology.*
10. Barry Commoner, *The Closing Circle.*
11. Numerous works are available on the impact of the Aswan
Dam. See for example, the following articles in Farvar and
Milton: Carl J. George, "The Role of the Aswan High Dam
in Changing the Fisheries of Southeastern Mediterranean";
M. Kassas, "Impact of River Control Schemes on the Shore-
lines of the Nile Delta"; E. Barton Worthington, "The Nile
Catchment—Technological Change and Aquatic Biology."
12. Some Chinese engineers had argued strongly for postponing
construction at San-men Gorge, but they were overwhelmed
by those who supported early construction at this site. The
case of Huang Wan-li, a professor of hydraulic engineering
at Tsinghua University was the most widely publicized.
Huang argued against the over-optimistic view that heavy
investment at San-men Gorge would be a panacea for Yellow River water management ills. His opinions were an-
swered by a battery of articles collected in *CKSL*, September
1957. See Huang's articles, "A Small Voice among the
Flowers (Hua-tsung Hsiao-yu)," *JMJP*, June 19, 1957, and
"An Opinion on the Planning Method for the San-men
Gorge Reservoir (Tui-yu Huang-ho San-men-hsia Shui-k'u
Hsien-hsing Kuei-hua Fang-fa Ti Yi-chien)," *CKSL*, No. 8
(August 1958), pp. 26–29.
13. Liu Shen, "Tentative Views on Some Basic Questions in Our
Country's Water Conservancy Construction," *Economic Re-*

search (*Ching-chi Yen-chiu*), No. 6 (June 20, 1965), *SCMM*, No. 483 (August 9, 1965), pp. 24–32.

14. Chao Ming-fu, "Provide Stronger Leadership."
15. NCNA, Peking, December 21, 1957.
16. Donald J. Munro, *The Concept of Man in Early China.*
17. NCNA, San-men Gorge, April 19, 1957, "Cooperatives Join in Subduing Yellow River," *SCMP*, No. 1516 (April 25, 1957), pp. 9–10.
18. Chao Ming-fu, "Provide Stronger Leadership."
19. Liu Yeh, "The Lifeline of Production."
20. *JMJP*, October 16, 1965, "Water Conservation by Depending on the 500 Million Peasants," *SCMP*, No. 3571 (November 3, 1965), pp. 7–10.
21. Chang Tze-lin, "Water Conservancy and Agricultural Production," *JMJP*, August 29, 1963, *SCMP*, No. 3065 (September 23, 1963), pp. 10–16.
22. Rhoads Murphey, "Man and Nature in China," *Modern Asian Studies*, Vol. I, No. 4 (1967), pp. 313–333.
23. Ch'in Fu, *Plans for Regulating Rivers.*
24. Li Yi-chih, "Discussion of Basic Management Method for the Yellow River," in *LYCCC.*
25. Marx, *The Machine in the Garden*; Roderick Nash, *Wilderness and the American Mind.*
26. *JMJP*, September 8, 1957, "An Important Task in the Reconstruction of Nature," *SCMP*, No. 1614 (September 20, 1957), pp. 22–25; Chang Chung-liang, "To Reform Nature It Is Necessary to Reform Thought First," *JMJP*, May 17, 1958, *CB*, No. 509 (June 10, 1958), pp. 27–30; *JMJP*, April 21, 1966, "Unlimited Creative Power of the Great Masses of the People," *SCMP*, No. 3686 (April 28, 1966), pp. 10–12.
27. A report on the flood emergency program in 1958 was entitled, "Decisive Victory Won in Battle against Yellow River Torrent," for example, (NCNA, Peking, July 25, 1958, *SCMP*, No. 1821, July 25, 1958), while an editorial in *People's Daily* on the soil conservation program was entitled, "An Important Task in the Reconstruction of Nature" (*JMJP*, September 8, 1957, *SCMP*, No. 1614 (September 20, 1957), pp. 22–25).
28. *JMJP*, March 12, 1966, "Planting of Trees Must be Preceded by

Nurturing of People," *SCMP*, 3686 (April 28, 1966), pp. 16–18.

29. White, Jr., "Historical Roots"; Moncrief, "Cultural Basis."
30. Chang Chung-liang, "To Reform Nature."
31. Revolutionary Mass Criticism Group of the Ministry of Forestry, "In Refutation of the View that 'Large Populations Will Destroy Forests,' " *JMJP*, February 25, 1970.
32. NCNA, Huhehot, October 10, 1961.
33. *JMJP*, May 28, 1971; see also Ma Szu-fu, "Cleverly Use Nature's Water at Ta-yu-chang (Ch'iao-ch'u T'ien-shui Ta-yu-chang)," *THJP* Tsinan, October 3, 1957.
34. *JMJP*, October 16, 1974b.
35. "Geography of China: Birds and Animals," *China Reconstructs*, Vol. 21, No. 7 (July 1972), pp. 39–40.
36. Mencius, Book 3, Part 1, *Chuan* IV:7. *The Chinese Classics*, translated by James Legge. Vol. 2, The Works of Mencius, pp. 250–251.

Bibliography

English Language Materials

BOOKS AND MONOGRAPHS

Barnett, A. Doak. *Cadres, Bureaucracy, and Political Power in Communist China*. New York: Columbia University Press, 1967.

Bennett, Adrian Arthur. *John Fryer: The Introduction of Western Science and Technology into Nineteenth Century China*. Harvard East Asian Monograph Series, No. 24. Cambridge, Mass.: Harvard University Press, 1967.

Chang Kuang-chih. *The Archaeology of Ancient China*. New Haven: Yale University Press, 1963.

Chang Shichin. *General Description of the Yellow River Basin*. YRPS, No. 2. Nanking, 1946.

Chi Ch'ao-ting. *Key Economic Areas in Chinese History*. London: George Allen and Unwin, 1936.

Commoner, Barry. *The Closing Circle*. New York: Alfred A. Knopf, 1971.

Cotton, John S. *Preliminary Report on the Yellow River Project*. YRPS, No. 11. Nanking, 1947.

El-Kammash, Magdi M. *Economic Development and Planning in Egypt*. New York: Praeger, 1968.

Farvar, M. Taghi, and Milton, John P., eds. *The Careless Technology*. Garden City, New York: The Natural History Press, 1972.

Goode, A. T., et al. *Report of the Committee of Experts on Hydraulic and Road Questions in China*. Geneva: League of Nations, 1936.

Haas, Robert. *Report of the Secretary of the Council Committee on His Mission in China*. Geneva: Council Committee on Technical Cooperation between the League of Nations and China, 1935.

Huang, W. H. *Soil and Water Conservation of the Yellow River Basin. YRPS*, No. 5. Nanking, 1946.

International Bank for Reconstruction and Development. *The Economic Development of Iraq.* Baltimore: Johns Hopkins University Press, 1952.

Langley, Kathleen M. *The Industrialization of Iraq.* Cambridge, Mass.: Harvard University Press, 1962.

Lattimore, Owen. *Inner Asian Frontiers of China.* New York: American Geographical Society, 1940.

Lee, J. S. *The Geology of China.* London: Thomas Murby and Co., 1939.

Mabro, Robert. *The Egyptian Economy 1952–1972.* Oxford: Clarendon Press, 1974.

Mallory, Walter H. *China: Land of Famine.* American Geographical Society Special Publication No. 6. New York, 1926.

Marx, Leo. *The Machine in the Garden.* New York: Oxford University Press, 1967.

Meng, S. M. *The Tsungli Yamen: Its Organization and Functions.* Harvard East Asian Monograph Series, No. 13. Cambridge, Mass.: Harvard University Press, 1962.

Michel, Aloys Arthur. *The Indus Rivers: A Study of the Effects of Partition.* New Haven: Yale University Press, 1967.

Mumford, Lewis. *Technics and Civilization.* New York: Harcourt, Brace and World, 1934.

Munro, Donald J. *The Concept of Man in Early China.* Stanford, California: Stanford University Press, 1969.

Nash, Roderick. *Wilderness and the American Mind.* New Haven: Yale University Press, 1967.

Nathan, Andrew James. *A History of the China International Famine Relief Commission.* Harvard East Asian Monograph Series, No. 17. Cambridge, Mass.: Harvard University Press, 1965.

Needham, Joseph. *Mechanical Engineering.* Vol. 4, No. 2, of *Science and Civilization in China.* Cambridge: Cambridge University Press, 1965.

———. *Civil Engineering and Nautics.* Vol. 4, No. 3, of *Science and Civilization in China.* Cambridge: Cambridge University Press, 1971.

Rajchman, L. *Report of the Technical Agent of the Council on His Mission in China.* Geneva: Council Committee on Technical Cooperation between the League of Nations and China, 1934.

Reybold, Eugene; Growden, James P.; and Savage, John L. *Preliminary Report on the Yellow River Project. YRPS*, No. 10. Nanking, 1947.

Shen Yi. *Purpose and Scope of the Yellow River Project Studies. YRPS*, No. 1. Nanking, 1947.

Teng Tse-hui. *Report on the Multiple Purpose Plan for Permanently Controlling the Yellow River and Exploiting Its Water Resources.* Peking: Foreign Languages Press, 1955.

Thorp, James. *Geography of the Soils of China.* Nanking: The National Geological Survey of China, 1936.

Ting, S., et al. *Geology and Soils of the Yellow River Basin. YRPS*, No. 3. Nanking, 1946.

Todd, David Keith. *The Water Encyclopedia.* Port Washington, New York: The Water Information Center, 1970.

Todd, O. J. *Two Decades in China.* Peking: Association of Chinese and American Engineers, 1938. Reprint. Taipei: Ch'eng-wen Publishing Co., 1971.

White, Gilbert. *Strategies of American Water Management.* Ann Arbor: University of Michigan Press, 1969.

Wittfogel, Karl A. *Oriental Despotism.* New Haven: Yale University Press, 1957.

Wu Yuan-li. *The Spatial Economy of Communist China.* New York: Praeger, 1967.

Zaccone, June Mary. "Some Aspects of Surplus Labor, Water Control, and Planning in China, 1949–1960." Ph.D. dissertation, University of North Carolina, 1963.

Zaychikov, V. T., ed. *The Physical Geography of China (Fizicheskaya Kitaya).* Moscow, 1964. Translated by Joint Publications Research Service (JPRS). Washington, D.C.: U.S. Department of Commerce, 1965.

ARTICLES

Bishop, Carl Whiting. "The Beginnings of North and South in China." *Pacific Affairs*, Vol. 7, No. 3 (September 1934), pp. 297–325.

Brown, Leonard R. "Are River Basins the Best Framework for Water Resources Planning, Development, and Management?" *Water Resources Bulletin*, Vol. 8, No. 2 (April 1972), pp. 401–403.

Chang Kuei-sheng. "Nuclei Formation of Communist China's Iron and Steel Industry." *Annals of the Association of American Geographers,* Vol. 60, No. 2 (June 1970), pp. 257–285.

Chen Cheng-siang. "Population Growth and Urbanization in China, 1953–1970." *Geographical Review,* Vol. 63, No. 1 (January 1973), pp. 534–548.

Chi Wen-shun. "Water Conservancy in Communist China." *China Quarterly,* No. 23 (July–September, 1965).

Freeman, John R. "Flood Problems in China." *Transactions of the American Society of Civil Engineers,* Vol. 75 (1922), pp. 1405–1460.

"Geography of China: Birds and Animals." *China Reconstructs,* Vol. 21, No. 7 (July 1972), pp. 39–40.

Ho P'ing-ti. "The Loess and the Origin of Chinese Agriculture." *American Historical Review,* Vol. 75, No. 1 (October 1969), pp. 1–36.

Hu Ch'ang-tu. "The Yellow River Administration in the Ch'ing Dynasty." *Far East Review,* Vol. 14 (1955), pp. 505–513.

Lowdermilk, W. C. "Forestry in Denuded China." *Annals of the American Academy of Political and Social Science,* Vol. 152 (November 1930), pp. 127–141.

———. "Man-Made Deserts." *Pacific Affairs,* Vol. 8, No. 4 (December 1935), pp. 409–419.

Meisner, Maurice. "The Despotism of Concepts: Wittfogel and Marx on China." *China Quarterly,* No. 16 (October–December 1963), pp. 99–111.

Moncrief, Lewis W. "The Cultural Basis for Our Environmental Crisis." *Science,* Vol. 170, No. 3957 (October 30, 1970), pp. 508–512.

Murphey, Rhoads. "Man and Nature in China." *Modern Asian Studies,* Vol. 1, No. 4 (1967), pp. 313–333.

Schwarz, Henry G. "Chinese Migration to Northwest China and Inner Mongolia, 1949–1959." *China Quarterly,* No. 16 (October–December 1963), pp. 62–74.

Shen Su-ju. "At the Great Bend of the Yellow River." *China Reconstructs* (July 1964), pp. 32–35.

Todd, O. J. "Discussion of Saratsi Irrigation Project." *Journal of the Association of Chinese and American Engineers* (May–June 1935).

———. "Repairing the Yellow River Break, Kung Chia Ko, Shantung." *Far East Review,* Vol. 19, (July 1923), pp. 468–481.

———. "The Yellow River Breaks of 1935." *Far East Review,* Vol.

32 (October 1936), pp. 454–461.

———. "The Yellow River Reharnessed." *Geographical Review,* Vol. 39, No. 1 (January 1949), pp. 38–56.

Todd, O. J., and Eliassen, Siguard. "The Yellow River Problem." *Transactions of the American Society of Civil Engineers,* Vol. 105 (1940), pp. 346–453.

Vitvitskiy, G. N. "Climate." In V. T. Zaychikov, ed., *The Physical Geography of China (Fizicheskaya Kitaya).* Moscow, 1964. Translated by Joint Publications Research Service (JPRS). Washington, D.C.: U.S. Department of Commerce, 1965.

"Water for the Loess Highlands." *China Reconstructs,* August 1970, p. 22.

White, Lynn, Jr. "The Historical Roots of Our Ecological Crisis." *Science,* Vol. 155 (1967), pp. 1203–1207.

Wu Ta-k'un. "An Interpretation of Chinese Economic History." *Past and Present,* Vol. 1, No. 1 (February 1952), pp. 1–12.

Zaychikov, V. T. "Inland Waters." In idem, ed., *The Physical Geography of China (Fizicheskaya Kitaya),* Moscow, 1964. Translated by Joint Publications Research Service (JPRS). Washington, D.C.: U.S. Department of Commerce, 1965.

Chinese and Japanese Language Materials

BOOKS

Ajia no Yume (Dream of Asia: Water Resources of China). Tokyo, 1964. Translated by Joint Publications Research Service (JPRS No. 32, 681).

Archaeological Research Institute, Chinese Academy of Sciences (Chung-kuo K'o-hsueh-yuan, K'ao-ku Yen-chiu-so). "Remnants of the San-men Gorge Transport Canal (San-men-hsia Ts'ao-yun I-chi)." *Collected Chinese Archaeological Field Reports (Chung-kuo T'ien K'ao-ku Pao-kao Chi).* Peking, 1959.

Chang Han-ying, ed. *Report on an Investigation of the Upper and Middle Reaches of the Yellow River (Huang-ho Shang-chung-yu K'ao-ch'a Pao-kao).* YRWCC, 1947.

———. *Sketch of Historical Yellow River Management Methods (Li-tai Chih-ho Fang-lueh Shu-yao).* Shanghai: Commercial Press, 1944.

Ch'in Fu. *Plans for Regulating Rivers (Chih-ho Fang-lueh).* 1799.

Combined Investigation Team on Soil and Water Conservation in the Middle Yellow River, Chinese Academy of Sciences. *Soil and Water Conservation Handbook (Shui-t'u Pao-ch'ih Shou-ts'e)*. Peking: Science Publishing Co., 1959.

Department of Higher Education, Ministry of Education. *A List of Universities, Independent Colleges, and Technical Schools in China, May, 1935*. Nanking, 1935.

Li Yi-chih. *The Collected Works of Li Yi-chih (Li Yi-chih Ch'uan-chi)*. Taipei: China Collections Committee (Chung-hua Ts'ung-shu Wei-yuan-hui), 1956.

Mencius, Book 3, Part 1, *Chuan* IV:7. In *The Chinese Classics*, translated by James Legge, Vol. 2, *The Works of Mencius*, pp. 250–251. Oxford: Clarendon Press, 1893–1895.

Ministry of Water Conservancy (Shui-li Pu). *Water Conservancy Administration (Shui-li Hsing-cheng)*. 1947.

Shen Yi, ed. *Collected Essays on the Yellow River Problem (Huang-ho Wen-t'i T'ao-lun Chi)*. Taipei: Taiwan Commercial Press, 1971.

Sung Hsi-shang. *Biography of Li Yi-chih (Li Yi-chih Ti Sheng-p'ing)*. Taipei: China Collections Committee, 1964.

Ts'en Chung-mien. *History of the Changes of the Yellow River (Huang-ho Pien-ch'ien Shih)*. Peking: People's Publishing Co., 1957.

Yellow River Water Conservancy Commission. *Yellow River Chronicle (Huang-ho Chih)*. 1935.

———. *Yellow River Hydrology (Huang-ho Shui-wen)*. 1945.

SIGNED ARTICLES

Chang Chung-liang. "To Reform Nature It Is Necessary to Reform Thought First." *JMJP*, May 17, 1958; *CB*, No. 509, June 10, 1958, pp. 27–30.

Chang Tze-lin. "Water Conservancy and Agricultural Production." *JMJP*, August 29, 1963; *SCMP*, No. 3065, September 23, 1963, pp. 10–16.

Chao Ming-fu. "A Discourse on Soil Conservancy in the Yellow River Basin." *Red Flag*, No. 21, November 21, 1962; *SCMM*, No. 341, November 26, 1962, pp. 29–37.

———. "Provide Stronger Leadership and Rely on the Masses in Bringing About a New Upsurge in the Water and Soil Conservation Work in the Middle Reaches of the Yellow River."

SLYTL, No. 24, December 20, 1963; *SCMM*, No. 403, February 10, 1964, pp. 6–11.

———. "Water and Soil Conservation along the Middle Reaches of the Yellow River." *SLYTL*, No. 13, July 5, 1963; *SCMM*, No. 380, September 3, 1963, pp. 1–8.

Ch'en Chao-ping. "What We Have Learned from the Soviet Experts (Wo-men Ts'ung Su-lien Chuan-chia-men Hsueh-hsi So Shem-me)." *CKSL*, No. 11, November 1957, pp. 8–9.

Ch'in Hsu-lun. "The Great Bend of the Yellow River Today (Chin-jih ti Huang-ho Ho-t'ao)." *HKTKP*, June 7, 1955.

Chun Chien. "The Outstanding Contribution of the Soviet Advisors in the San-men Gorge Project (Su-lien Chuan-chia Tui San-men-hsia Shui-li Shu-niu Kung-ch'eng ti Cho-yueh Kung-hsien)." *JMJP*, April 15, 1957.

———. "Work Completed on Shih-t'ou-chuang Overflow Weir (Shih-t'ou-chuang Yi-hung-yen Wan-kung)." *JMJP*, September 10, 1951.

Fang Cheng-san. "An Exploration into the Question of Water and Soil Conservation on the Loess Highland in the Northwest." *JMJP*, May 24, 1965; *SCMP*, No. 3480, June 18, 1965, pp. 7–11.

Feng Ching-lan. "Special Features of the Physiographic Regions of the Yellow River Basin (Huang-ho Liu-yu Ti-hsing Fen-ch'iu Ti T'e-tien)." *KMJP*, August 1, 1955.

Han Fang. "Yellow River Water and Hua-yuan-k'ou (Huang-ho Shui Yu Hua-yuan-k'ou)." *YCWP*, August 25, 1965.

Hsia, Kai J. and Lee, C. S. "A Study of the Alluvial Morphology of the Lower Wei Basin (Wei-ho Hsia-yu Chung-chi Hsing-t'ai ti Yen-chiu)." *TLHP*, Vol. 29, No. 3 (September 1963), pp. 207–218.

Hsieh Shih-yen and Chang Chih-ch'u. "A Visit to Ch'ing-t'ung Gorge (Yao-fang Ch'ing-t'ung-hsia)." *JMJP*, January 11, 1960.

Huang Li. "Can the Yellow River Irrigate the Land (Huang-ho Shui K'o-pu-k'o-yi Jao-ti)." *YCWP*, July 12, 1965.

Huang Wan-li. "An Opinion on the Planning Method for the San-men Gorge Reservoir (Tui-yu Huang-ho San-men-hsia Shui-k'u Hsien-hsing Kuei-hua Fang-fa Ti Yi-chien)." *CKSL*, No. 8, August, 1958, pp. 26–29.

———. "A Small Voice among the Flowers (Hua-tsung Hsiao-yu)." *JMJP*, June 19, 1957.

Koroliev, A. A. "The Key Yellow River San-men Gorge Water Con-

servancy Project (Huang-ho San-men-hsia Shui-li Shu-niu)."
   CKSL, No. 15, March, 1957, pp. 5–7.
————. "Problems Concerning Future Development of China's
   Water Conservancy Work (Kuan-yu Chung-kuo Shui-li Shih-
   yeh Chin-hou Fa-chan ti Mou-hsieh Wen-t'i)." *SLYTL*, No.
   4, April, 1959, pp. 19–32; *ECMM*, No. 168, May 18, 1959, pp.
   25–36.
Kuo Ching-hui. "The Silt of the Yellow River and Its Erosive Action
   (Huang-ho ti Ni-sha Chi Ch'i Ch'in-shih Tzo-yung)." *TLCS*,
   August, 1956.
Lan K'o-chen. "On the Northward Transfer of Southern Waters (Lun
   Nan-shui Pei-tiao)." *TLCS*, April, 1959, pp. 145–146.
Lei Chien. "Desolate San-men Gorge (Leng-lo-le ti San-men hsia)."
   HTJP, July 27 and 28, 1963.
Li Fu-tu. "An Important Way to Solve the Problem of Silt and Sand
   in the Yellow River." *JMJP*, January 26, 1966.
Li Yi-chih. "Discussion of the Basic Management Method for the
   Yellow River (Huang-ho Chih Ken-pen Chih-fa Shang-choi)."
   LYCCC, pp. 401–423.
————. "General Overview of a Yellow River Management Plan
   (Huang-ho Chih-pen Chi-hua Kai-yao Su-mu)." *LYCCC*, pp.
   441–450.
————. "Opinion on Yellow River Management (Chih-huang Yi-
   chien)." *LYCCC*, pp. 463–465.
————. "An Opinion on Yellow River Management (Kuan-yu
   Huang-ho Chih-tao Chih Yi-chien)." *LYCCC*, pp. 465–467.
————. "Outline for Yellow River Management Work (Chih-li
   Huang-ho Kung-tso Kang-yao)." *LYCCC*, pp. 458–463.
————. "Work Plan for the Yellow River Water Conservancy Com-
   mission (Huang-ho Shui-li Wei-yuan-hui Kung-tso Chi-
   hua)." *LYCCC*, pp. 450–458.
Liu Shen. "Tentative Views on Some Basic Questions in Our Coun-
   try's Water Conservancy Construction." *Economic Research*
   (*Ching-chi Yen-chiu*), No. 6, June 20, 1965; *SCMM*, No. 483,
   August 9, 1965, pp. 24–32.
Liu Te-jen. "The Assistance of Soviet Experts in Project Manage-
   ment Work (Su-lien Chuan-chia Tui Kung-ch'eng Kuan-li
   Kung-tso ti Pang-chu)." *CKSL*, No. 11, November, 1957, pp.
   6–7.
Liu Tze-hou. "The Assistance of Soviet Experts in the San-men
   Gorge Project (Su-lien Chuan-chia Tui San-men-hsia Shui-li

Kung-ch'eng ti Pang-chu)." *Wen-hui Pao* (Shanghai), November 6, 1957.

Liu Yeh. "The Lifeline of Production in Mountainous Areas." *JMJP*, January 18, 1962; *SCMP*, No. 2674, February 8, 1962, pp. 4–15.

Ma Szu-fu. "Cleverly Use Nature's Water at Ta-yu-chang (Ch'iao-ch'u T'ien-shui Ta-yu-chang)." *THJP*, Tsinan, October 3, 1957.

Meng Chin-chih. "The New Face of the Yellow River (Huang-ho ti Hsin Mien-mao)." *HKWHP*, August 23, 1966.

Meng Lin. "More on the Matter of the Communists Blasting the San-men Gorge Reservoir (Tsai T'an Chung-kung Tza-hui San-men-hsia Shui-k'u Shih-chien)." *CCNW*, March 16, 1965.

———. "News of the Communists Blasting San-men Gorge Reservoir (Chung-kung Tza-hui San-men-hsia Shui-k'u Chi-wen)." *CCNW*, December 16, 1964.

Revolutionary Mass Criticism Group of the Ministry of Forestry. "In Refutation of the View That 'Large Populations Will Destroy Forests.' " *JMJP*, February 25, 1970.

Shao Tang. "Overflow Weir Project to Control the Yellow River (Chih-li Huang-ho ti Yi-hung-yen Kung-ch'eng)." *HKTKP*, July 11, 1954.

Shen Ts'ung-kang. "On the Planning Situation for the First Stage at San-men Gorge (San-men-hsia Ch'u-pu She-chi Ch'ing-kuang Chieh-shao)." *CKSL*, No. 19, July, 1957, pp. 11–15.

T'ang Shih. "What the Soviet Experts Have Contributed to China's Power Industry in the Past Eight Years." *People's Power Industry (Jen-min Tien-yeh)*, No. 31, November 5, 1957; *ECMM*, No. 116, January 27, 1958, pp. 34–38.

T'ao Tuan-shih. "The Routes Question in Diversion of the Han, Yellow, and Huai Rivers (Yin-han Chi-huang Chi-huai Lu-hsien Wen-t'i)." *JMJP*, June 4, 1956.

Tung Keng. "Winter Water Conservation Works Show Marked Results." *Current Events Handbook (Shih-Shih Shou-ts'e)*, No. 2, January 16, 1965; *SCMM*, No. 460, March 16, 1965, pp. 24–25.

Wang Ch'ien-kuang. "On the Question of the Regular Water Level for the San-men Gorge Hydropower Station (Kuan-yu San-men-hsia Shui-tien-chan Cheng-ch'ang Kao-shui Wen-t'i)." *SLFT*, 1957.

Wang Hua-yun. "The Great Rationality of Transferring Southern
    Waters Northward (Nan-shui Pei-tiao ti Hsiung-wei Li-
    hsiang)." *Red Flag (Hung Ch'i)*, September 1, 1959, pp. 36–
    44.

————. "The Great Significance of the Key Project at San-men
    Gorge (Huang-ho San-men-hsia Shui-li Shu-niu Kung-ch'eng
    ti Chung-yao Yi-yi)." *CKSL*, No. 15, March, 1957, pp. 1–4.

Yang K'uan. "Achievements of Hydraulic Engineering in the War-
    ring States Period (Chan-kuo Shih-tai Shui-li Kung-ch'eng
    ti Ch'eng-chiu)." In Li Kuang-pi and Ch'ien Chun-yeh, eds.,
    *Essays on Chinese Scientific Discoveries and Scientists
    (Chung-kuo K'e-hsueh Chi-shu Jen-wu Lun-chi)*. Peking,
    1955.

Yeh Yung-yi. "Some Special Points on Solving the Flood Problem in
    the Yellow River Management Plan (Chih-li Huang-ho Kuei-
    hua Chung Chieh-chueh Fang-hung Wen-t'i Chi-ke T'e-
    tien)." *KMJP*, August 4 , 1955.

UNSIGNED ARTICLES

HCJP

1958. March 22. "Ten Thousand Workers Take Part in River Dredg-
    ing Project (Wan-jen Ts'an-chia Wa-ho Kung-ch'eng)."

HKWHP

1959. May 20. "The Northward Transfer of Southern Waters Has
    Two Dragons Boring through Mountains and Going over
    Peaks to Bring Water (Nan-shui Pei-tiao Yu Shuang-lung
    Ch'uan-shan Yueh-ling Tsai Shui Lai)."

HTJP

1951. November 11. "Construction Completed on Huang-yang
    Gate Irrigation Project in the Ho-t'ao Region of Sui-yuan
    (Sui-yuan Ho-t'ao Huang-yang-cha Kuang-kai Kung-ch'eng
    Yi Ts'ao-ch'eng)."

JMJP

1955. March 31. "Huge Temporary Flood Prevention Project Built
    on Yellow River Lower Reaches (Huang-ho Hsia-yu Hsing-
    hsiu Chu-ta ti Lin-shih Fang-hung Kung-ch'eng)."

1956. May 12. "Spring Repair Project Completed on Yin-ch'uan Special District Canal (Yin-ch'uan Ch'uan-ch'u Wan-ch'eng Ch'u-tao Ch'un-hsiu Kung-ch'eng)."

1957. September 8. "An Important Task in the Reconstruction of Nature." *SCMP*, No. 1614, September 20, 1957, pp. 22–25.

————. September 25. "Raise an Enthusiastic Wave of Water Conservancy Construction for Agricultural Land." *SCMP*, No. 1626, October 8, 1957, pp. 16–18.

1958. April 9. "Construction Begun on Kang-li Yellow River Diversion Irrigation Project (Kang-li Yin-huang Kuang-kai Kung-ch'eng K'ai-kung)."

————. September 12. "CCP Central Committee Directive on Water Conservancy Work." *SCMP*, No. 1857, September 19, 1958, pp. 4–6.

1962. March 23. "Reclamation of Wasteland on Steep Slopes Must Be Restricted." *SCMP*, No. 2719, April 16, 1962, pp. 15–16.

1963. November 11. "Winter Reconstruction Projects Successively Begin Work on Lower Yellow River (Huang-ho Hsia-yu Tung-hsiu Kung-ch'eng Lu-hsu K'ai-kung)."

1964. November 11. "Build a Drainage Network, a Year Round Remedy to Waterlogging (Chien-chii P'ai-shui-kang Ch'uan-nien Wu Nei-lao)."

1965. May 23. "Li ch'eng Cleverly Uses Yellow River Benefits and Cleverly Eliminates Silt Water Disaster (Li ch'eng Ch'iao-yung Huang-ho Chih-li Ch'iao-pi Huang-shui Chih-hai)."

————. September 9. "National Conference on Water Conservation Stresses That Water Conservation Should Render Better Service for Increasing Agricultural Production." *SCMP*, No. 3435, April 8, 1965, pp. 11–13.

————. October 16. "Water Conservation by Depending on the 500 Million Peasants." *SCMP*, No. 3571, November 3, 1965, pp. 7–10.

1966. March 12. "Planting of Trees Must Be Preceded by Nurturing of People." *SCMP*, No. 3686, April 28, 1966, pp. 16–18.

————. April 21. "Unlimited Creative Power of the Great Masses of the People." *SCMP*, No. 3686, April 28, 1966, pp. 10–12.

1971. May 26. "Utilize Water and Sand Resources of the Yellow River, Build Stable-Yield and High-Yield Farm Fields." *SCMP*, No. 4914, June 9, 1971, pp. 117–122.

————. May 28. "Drawing Yellow River Water for Irrigation, Transforming Harm into Blessing." *SCMP*, No. 4918, June 15, 1971, pp. 53–63.

1974. September 16a. "Key Ch'ing-t'ung Gorge Water Conservancy Project Basically Completed (Ch'ing-t'ung-hsia Shui-li Shu-niu Kung-ch'eng Chi-pen Chien-ch'eng)."
———. September 16b. "Long Journey on the Yellow River (Huang-ho Wan-li Hsing)."
———. December 20a. "Reconstruction of Key San-men Gorge Water Conservancy Project Attains Initial Success (San-men-hsia Shui-li Shu-niu Kung-ch'eng Kai-chien Huo-te Ch'u-pu Ch'eng-kung)."
———. December 20b. "Understand the Yellow River, Rebuild the Yellow River (Jen-shih Huang-ho Kai-chien Huang-ho)."
1975. February 5. "Our Country's Biggest Hydroelectric Power Station—Liu-chia Gorge Hydroelectric Power Station—Victoriously Completed (Wo-kuo Tsui-ta-ti Shui-tien-chan—Liu-chia-hsia Shui-tien-chan—Sheng-li Chien-li Chien-ch'eng)."

KJJP

1962. January 5. "New Face of San-men Gorge (San-men-hsia Hsin-mao)."

KMJP

1957. January 11. "Soil Conservation Suffers on Middle and Downstream of Yellow River." *SCMP*, No. 1455, January 22, 1957, pp. 13–14.
1971. February 6. "Vigorously Make a Success of Water and Soil Conservation, Develop Agricultural Production in Mountain Areas." *SCMP*, No. 4841, February 19, 1971, pp. 134–141.

NMKJP

1956. July 28. "A Strong Yellow River Levee (Yi-tao Chien-ku ti Huang-ho T'i)."

SJP

1959. November 26. "Resettlement of People in the Area of the San-men Gorge Reservoir, Shensi, Commences." *SCMP*, No. 2173, June 11, 1960, pp. 16–18.

TKP

1958. September 14. "Brave People and the Bravery Canal (Ying-hsiung Jen-min Ying-hsing Ch'u)."

PRESS RELEASES

CKHW
1960. January 10. Tsinan. "Construction Begins on a Large-Scale Water Conservancy Project in Shantung (Shan-tung Yi-ko Ta-hsing Shui-li Kung-ch'eng Tung-kung Hsing-chien)."
1962. April 13. Kuangchou. "Yellow River Irrigation District Increases Spring Repairs (Huang-ho Kuang-ch'u Cha-fan Ch'un-hsiu)."

Kansu Provincial Radio

1974. September 24. "Kansu Completes Yellow River Hydroelectric Plant." *FBIS*, September 26, 1974.

NCNA

1953. January 7. Kaifeng. "Exact Source of Yellow River Thoroughly Explored." *SCMP*, No. 487, January 8, 1953, pp. 20–21.
———. April 6. Kaifeng. "Wheat Fields Irrigated by Yellow River in Honan." *SCMP*, No. 545, April 5–7, 1953, p. 31.
———. July 4. Kaifeng. "Yellow River Headquarters Set Up." *SCMP*, No. 604, July 7, 1953, p. 7.
1954. April 4. Peking. "Dikes along Yellow River Being Strengthened." *SCMP*, No. 783, April 7, 1954, p. 8.
———. May 28. Peking. "Fu Tso-yi Reviews Water Conservancy Work in China: Irrigation Aids Agriculture." *SCMP*, No. 821, June 3, 1954, pp. 29–30.
———. November 2. Lanchou. "Yellow River Soil Erosion Can Be Checked, Scientists Say." *SCMP*, No. 921, November 3, 1954, pp. 11–12.
1955. April 25. Chengchou. "Exhibition on Yellow River Control Opened in Chengchou." *SCMP*, No. 1036, April 28, 1955, p. 52.
———. July 22. Taiyuan. "Soil Erosion along the Yellow River in Shansi." *SCMP*, No. 1095, July 23, 1955, p. 41.

————. August 11. Peking. "Yellow River Control Work Reviewed." *SCMP*, No. 1108, August 12, 1955, pp. 15–16.

————. September 19. Lanchou. "New Irrigation Canal along Yellow River Reaches." *SCMP*, No. 1133, September 20, 1955, p. 53.

1956.   May 5. Huhehot. "Nine Thousand Volunteer Land Reclaimers Arrive in Inner Mongolia." *SCMP*, No. 1302, June 5, 1956, p. 24.

————. May 19. Chengchou. "Yellow River Diversion Scheme in Honan." *SCMP*, No. 1295, May 24, 1956, p. 25.

————. June 28. Peking. "Yellow River Water Crest Approaches Coast." *SCMP*, No. 1324, July 9, 1956, p. 5.

————. June 29. Chengchou. "Yellow River Water Diversion Project Completed." *SCMP*, No. 1324, July 9, 1956, pp. 14–15.

————. July 1. Chengchou. "Yellow River Flood Prevention Headquarters Issues Directive on Possible Arrival of Larger Floods." *SCMP*, No. 1329, July 10, 1956, pp. 18–19.

————. July 22. Sian. "Water and Soil Conservation Work in Shensi." *SCMP*, No. 1336, July 25, 1956.

————. August 8. Yenan. "Survey of Wu-ting River in Shensi Completed." *SCMP*, No. 1348, August 13, 1956.

————. October 25. Chengchou. "Hydrological Survey along Yellow River." *SCMP*, No. 1400, October 30, 1956, p. 16.

————. October 30. Chengchou. "Ching River Survey Completed." *SCMP*, No. 1403, November 2, 1956, p. 24.

1957.   March 29. San-men Gorge. "Construction of San-men Gorge Will Start Soon." *SCMP*, No. 1503, April 3, 1957, pp. 13–14.

————. April 15. San-men Gorge. "Machines Get into Action at San-men." *SCMP*, No. 1514, April 23, 1957, p. 4.

————. April 19. San-men Gorge. "Cooperatives Join in Subduing Yellow River." *SCMP*, No. 1516, April 25, 1957, pp. 9–10.

————. June 3. Chengchou. "Progress of Water and Soil Conservation Work along Upper and Middle Reaches of Yellow River Criticized." *SCMP*, No. 1552, June 18, 1957, pp. 23–24.

————. June 30. Shan-pa, Inner Mongolia. "Projects to Reconstruct Irrigation Channels along Pao-t'ou–Lanchou Railway in Inner Mongolia Completed." *SCMP*, No. 1569, July 15, 1957, p. 25.

————. December 21. Peking. "Yellow River Tributary Basin Curtailed." *SCMP*, No. 1680, December 30, 1957, pp. 7–8.

1958.   March 13. Peking. "San-men Gorge Project to Be Completed Ahead of Schedule." *SCMP*, No. 1735, March 20, 1958, pp. 7–8.

———. May 28. Chengchou. "Conference of Yellow River Anti-flood Headquarters Decides to Strive for Greatest Victory Possible." *SCMP*, No. 1790, June 12, 1958, pp. 18–19.

———. June 6. San-men Gorge. "San-men Gorge Project Going Ahead at Top Speed." *SCMP*, No. 1791, June 13, 1958, p. 13.

———. July 18. Chengchou. "River Fighters Guard Yellow River Dikes." *SCMP*, No. 1818, July 24, 1958, pp. 36–37.

———. July 20. Peking. "Yellow River Torrent Reaches Coastal Shantung Province." *SCMP*, No. 1819, July 25, 1958, pp. 9–10.

———. July 22. Tsinan. "Yellow River Water Diverted by Lake." *SCMP*, No. 1820, July 28, 1958, p. 19.

———. July 23. Chengchou. "Yellow River Crest Safely Passes Hazardous Section." *SCMP*, No. 1821, July 29, 1958, p. 15.

———. July 25. Peking. "Decisive Victory Won in Battle against Yellow River Torrent." *SCMP*, No. 1821, July 25, 1958.

———. August 5. Chengchou. "Meeting Plans to Strengthen Yellow River Projects." *SCMP*, No. 1831, August 13, 1958, pp. 37–38.

———. September 28. Lanchou. "Two New Multipurpose Projects along the Yellow River." *SCMP*, No. 1867, October 3, 1958, p. 32.

———. October 18. Chengchou. "Preliminary Survey of Yangtze–Yellow River Canal Completed." *SCMP*, No. 1830, October 23, 1958, p. 23.

———. October 21. Chengchou. "Another Yangtze–Yellow River Canal Planned." *SCMP*, No. 1881, October 24, 1958, pp. 25–26.

1959. January 11. Yin-ch'uan. "Irrigation Network Being Built in Southern Ninghsia." *SCMP*, No. 1934, January 15, 1959, p. 24.

———. January 12. Taiyuan. "Junks Sail on Mountainous Area in Shansi." *SCMP*, No. 1934, January 15, 1959, p. 22.

———. March 15. Chengchou. "China's Sorrow Now Serves Farmers." *SCMP*, No. 1975, March 18, 1959, pp. 36–37.

———. June 5. Yin-ch'uan. "Experiments to Improve Alkaline Soils in Northwest China." *SCMP*, No. 2032, June 11, 1959, pp. 24–25.

———. August 8. Chengchou. "New Stage in Yellow River Multiple Purpose Project." *SCMP*, No. 2075, August 13, 1959, pp. 25–26.

———. August 22. Yin-ch'uan. "First Stage of Big Reservoir Finished on Yellow River Tributary." *SCMP*, No. 2085, August 27, 1959, p. 30.

———. September 16. Chengchou. "Work Begins on Big Yellow River Hydroelectric Power Station." *SCMP*, No. 2100, September 22, 1959, p. 29.

———. September 22. Peking. "Mass Campaign to Control Soil Erosion to Be Launched in Yellow River Basin." *SCMP*, No. 2104, September 28, 1959, p. 30.

———. October 3. Chengchou. "China's Sorrow Is Being Tamed." *SCMP*, No. 2131, November 5, 1959, pp. 28–30.

———. November 17. Peking. "Yellow River Water and Soil Conservation Projects." *SCMP*, No. 2142, November 24, 1959, p. 7.

———. December 2. Chengchou. "Work Begins on Seventh Giant Yellow River Reservoir." *SCMP*, No. 2151, December 8, 1959, p. 40.

———. December 10. San-men Gorge. "Builders of San-men Gorge Dam on Yellow River Complete 1959 Target." *SCMP*, No. 2157, December 21, 1959, p. 36.

———. December 15. Tsinan. "Yellow River Dam in Lower Reaches Completed." *SCMP*, No. 2161, December 22, 1959, p. 31.

———. December 31. Yin-ch'uan. "New Yellow River Irrigation Canal in Northwest China." *SCMP*, No. 2171, January 7, 1960, p. 27.

1960. January 3. Tsinan. "Yellow River Dam with Sorghum Stalks and Earth." *SCMP*, No. 2172, January 8, 1960, p. 15.

———. January 23. Tsinan. "Another Dam Being Built on Lower Reaches of Yellow River." *SCMP*, No. 2186, February 1, 1960, p. 10.

———. February 27. Tsinan. "Construction of New Yellow River Harnessing Project." *SCMP*, No. 2208, March 3, 1960, p. 12.

1961. March 19. Chengchou. "Big Irrigation System Expanded." *SCMP*, No. 2463, March 24, 1961, p. 17.

———. October 10. Huhehot. "Deserts in Inner Mongolia Being Tamed." *SCMP*, No. 2599, October 17, 1961, pp. 19–20.

1962. January 3. Sian. "Extensive Antierosion Work in Yellow River Basin." *SCMP*, No. 2654, January 9, 1962, pp. 25–26.

———. February 9. Peking. "Water Conservancy Projects Built

along Yellow and Yangtze Rivers." *SCMP*, No. 2679, February 16, 1962, pp. 20–22.

———. November 17. Sian. "Electric Pumps Lift Water Seventy Meters in Northwest China." *SCMP*, No. 2864, November 21, 1962, p. 10.

1963. August 19. Taiyuan. "Antierosion Measures by People's Communes in Yellow River Province." *SCMP*, No. 3045, August 22, 1963, pp. 11–12.

———. September 9. Hong Kong. "New Face of the Seventy-Two Connected Lakes (Ch'i-shih-erh Lien-hu Hsin-mao)."

———. September 17. Hong Kong. "Ninghsia–Yellow River Irrigation District Builds Three New Electrified Drainage Stations (Ning-hsia Huang-ho Kuang-kai-ch'u Hsin-chien San-tso Tien-li Pai-shui-chan)."

———. October 11. Hong Kong. "Shensi-Ninghsia Canal Siphon Projects Progress Well (Ch'in-ning Ch'u-tao Hsi-kuan Kung-ch'eng Chih-liang Hao)."

———. November 22. Yin-ch'uan. "Antierosion Work Transforms Loess Hills in Northwest China." *SCMP*, No. 3109, November 29, 1963, p. 13.

1964. May 10. Hong Kong. "Ninghsia–Yellow River Irrigation District Builds a New Drainage Project (Ning-hsia Yin-huang Kai-ch'u Hsin-chien Yi-p'i P'ai-shui Kung-ch'eng)."

———. June 30. Hong Kong. "Ninghsia–Yellow River Irrigation District Builds a Permanent Flood Prevention Levee (Ning-hsia Yin-huang Kuang-ch'u Hsiu-chien Yung-chiu-hsing Fan-hung Kung-ch'eng)."

———. October 13. Sian. "Antierosion Work in Yellow River Basin." *SCMP*, No. 3321, October 21, 1964, p. 18.

1965. June 15. Sian. "Massive Antierosion Project in Middle Reaches of Yellow River." *SCMP*, No. 3480, June 18, 1965, p. 18.

———. September 10. Sian. "Northwest China Exploits Underground Water for Irrigation." *SCMP*, No. 3537, September 15, 1965, p. 17.

———. October 22. Chengchou. "China's Scientists Work to Increase Soil Fertility in Yellow River Valley." *SCMP*, No. 3588, December 1, 1965, p. 19.

———. December 1. Peking. "People Make Semibarren Land in Northwest China Yield Stable Harvests." *SCMP*, No. 3591, December 6, 1965, p. 21.

1966. April 24. Yin-ch'uan. "Chinese Desert Control Workers Find

New Ways of Improving Extending Plant Cover." *SCMP*, No.
3686, April 28, 1966, p. 13.

————. October 31. Hong Kong. "Teng-k'ou Electric Pumping
Station in Inner Mongolia Is Completed and Begins Pumping (Nei-meng-ku Teng-k'ou Tien-li Yang-shui-chan Wan-
kung Ch'ou-shui)."

1970. April 16. Taiyuan. "Commune in Shansi Relies on Itself in
Water Conservancy Construction." *SCMP*, No. 4643, April
27, 1970, pp. 21–23.

1971. June 24. Peking. "Achievements in Using Yellow River Water
for Irrigation."*SCMP*, No. 4929, July 2, 1971, pp. 211–214.

————. October 31. Peking. "Mass Efforts at Soil Conservation
Help to Control Yellow River." *SCMP*, No. 5011, November
10, 1971, pp. 116–118.

**REPORTS AND PLANS OF THE YELLOW RIVER WATER
CONSERVANCY COUNCIL**

The Water Resources Planning Commission (WRPC) of the Ministry of Economics, Taipei, holds a large number of unpublished
plans and reports from experiment stations and other units working in the middle Yellow River basin prior to 1949. The following
materials, utilized for Chapter 3 of this book, are listed by WRPC
catalogue number.

59. "Summary of Materials on the Yellow River Undertaking
(Huang-ho Shui-li Shih-yeh Tz'u-liao Chi-yao)." February
1947.

71. "Postwar Five Year Construction Plan for the Yellow River
Commission (Huang-ho Shui-li Wei-yuan-hui Chan-hou Wu-
nien Chien-she Chi-hua)." August 1944.

76. "Water and Soil Conservation Plans and Investigation Reports of the Yellow River Commission (Huang-ho Shui-li
Wei-yuan-hui Shui-t'u Pao-ch'ih Chi-hua Chi Shih-ch'a Pao-
kao)."

   A. "Work Plans and General Calculations for Water and
   Soil Conservation Experiments in 1942."

   B. "Distribution Allotments of Water and Soil Conservation Work in 1945."

   C. "Work Plan for Soil and Water Conservation in 1943."

   D. "Work Plan for the Upstream Engineering Office of

Soil and Water Conservation Experiment District in 1945."

90. "Revised Plan for Soil and Water Conservation Experimental Work in 1944 (Hsiu-cheng San-shih-san Nien-tu Shui-t'u Pao-chi-ih Shih-hsien Kung-tso Chi-hua)." July 1944.

100. Yen Hsi-shan. "Draft Plan for Engineering in Shansi (Shan-hsi Kung-ch'eng Chi-hua)." October 1928.

106. "Plan Outline for the Fen River Hydropower and Irrigation Project in Shansi (Shan-hsi Fen-ho Shui-tien Kuang-kai Kung-ch'eng Chi-hua Kang-yao)."

112. "Shensi Province Water Conservancy Bureau, Statistics on Recently Completed New Irrigation and Water Conservancy in Shensi Province (Shen-hsi Hsin-hsing Kuang-kai Shui-li T'ung-chi)." October 1943.

120. "1947 Work Plan and General Budget of the Soil and Water Conservation Experiment District of the Upstream Engineering Office of the YRWCC (Huang-ho Shui-li Wei-yuan-hui Shang-yu Kung-ch'eng-ch'u Shui-t'u Pao-ch'ih Shih-hsien-ch'u San-shih-liu Nien-tu Kung-ch'eng Chi-hua Chi Kai-suan)."

121. "Preliminary Plan for Drainage and Irrigation in Chiao, Yu, Lin, and Yung Counties in the Alkaline Area of Shansi Province (Shan-hsi-sheng Chiao, Yu, Lin, Yung Hsien Chien-ti P'ai-shui Kuang-kai Kung-ch'eng Ch'u-pu Chi-hua)."

122. "Summary Work Plan for Soil Conservation in 1944 (San-shih-san Nien-tu Shui-t'u Pao-ch'ih Kung-tso Chi-hua Kang-yao)." December 1943.

134. Sui-yuan Province Water Conservancy Bureau. "Five Year Plan for Irrigation in Sui-yuan (Sui-yuan-sheng Kuang-kai Shih-yeh Wu-nien Chi-hua)."

151. Soil Conservation Experiment District of the Upstream Engineering Office of the YRWCC. "Plans for the Nan-chai Liu-yu Earth Dam (Nan-chai Liu-yu T'u-pa Kung-ch'eng Chi-hua-shu)."

152. A. "Draft Articles of Organization of the Lung-tung Soil and Water Conservation District (Lung-tung Shui-t'u Pao-ch'ih Shih-hsien-ch'u Tzu-chih Chang-cheng Ts'ao-an)."

B. "1943 Work Plan of the Lung-tung Soil Conservation Experiment Station (Lung-tung Shui-t'u Pao-ch'ih Shih-hsien-ch'u San-shih-erh-nien Kung-tso chi-hua)."

C. "1943 Work Plan for the Kuan-chung Soil Conservation

District (Kuan-chung Shui-t'u Pao-ch'ih Shih-hsien-ch'u San-shih-erh-nien Kung-tso chi-hua)."

153. A. "Work Plan of the Kuan-chung Soil Conservation Experiment District of the Upper Basin Forestry and Reclamation Project Office of the YRWCC."
    B. Shensi Agricultural Improvement Office, and Northwest Agricultural College. "Outline of the Method for Cooperatively Implementing the Kuan-chung District Soil and Water Conservation Plan."

157. "Investigation Report on 1943 Work of the Soil and Water Conservation Experiment District in the Lung-nan Area (Lung-nan-ch'u Shui-t'u Pao-ch'ih Shih-hsien-ch'u San-shih-erh Nien-tu Kung-tso Chien-t'ao Pao-kao)." March 1944.

178. "Practical Plan for Water and Soil Conservation Work in the Yellow River Basin (Huang-ho Liu-yu Shui-t'u Pao-ch'ih Kung-tso Shih-hsien Chi-hua)." July 1943.

Index

www.ingramcontent.com/pod-product-compliance
Ingram Content Group UK Ltd.
Pitfield, Milton Keynes, MK11 3LW, UK
UKHW041006050325
455862UK00002B/167